RECONNECT

Reestablishing Nature's Truth in Western Civilization

Daniel W. Lake

Published by Piscataqua Press

An imprint of RiverRun Bookstore, Inc.

32 Daniel Street

Portsmouth, NH 03801

www.riverrunbookstore.com

www.piscataquapress.com

ISBN: 978-1-950381-69-2

Printed in the United States of America

To Ethan and Madeleine:

A light to guide you along the home path.

The Big Idea

The root cause behind the multiple crises we are experiencing is faulty information.

False information is breaking multiple connections, facilitating both social and ecosystemic disintegrations. This false information was built, perhaps unintentionally, into the very foundation of Western Civilization and takes the form of **3-D's: Dualism, Dominionism, and Death.** Since these are religious non-truths, they must be replaced by religious truths. I use the term "religion" in its literal sense = to reconnect. If man and earth are to move forward in constructive partnership, each non-truth must be replaced by nature's truth—truth based on facts and verified by science. The purpose of this book is to outline understandings that facilitate renewed systemic integrity and **reconnect** us within a re-enchanted world.

Divided from earth, divided from our fellows, divided within ourselves, we constructed a **dominionist** narrative that allowed us to ravish the fertile land, commit genocides, become an agent of extinction, and a cancer that earth may soon remove for her own survival. Like the prodigal son who squandered his inheritance in wasteful living that would have caused his **death**, we are now starting to think of returning home again. Little voices are calling to us; memories suppressed for eons are tugging at our subconscious minds. "It's time to come home," they are saying, "time to return to the greater earth community and the nourishing relationships you once knew before word and thought became twisted. It's time to rejoin family, sit by the well-springs, and reconnect again."

Contents

Acknowledgments

No man is an island and we all stand on the shoulders of others. I am humbly grateful to all who have contributed to this book and its message. First of all is my wife of 52 years, Joyce, who has proofread this book many times. You have consistently believed in and encouraged me to write for a larger audience, to put my light on a "candlestick".

Additionally, I wish to thank Anne Vinnicombe for perusing much of the text and helping me state ideas clearly. Nancy Byrd worked with me through the entire text, making suggestions, catching both factual errors and unsubstantiated opinions. She has been a wonderful sounding board, unhesitating in challenging my thought, and the only polymath I know personally. I was fortunate to find an editor in Tom Holbrook of Piscataqua Press who was able to trim excess textual baggage, help clarify, and create a more natural flow of ideas. Working out of his River Run Bookstore, Tom is giving voice to hundreds of budding New England authors.

On the next ring of concentric influences I wish to acknowledge my indebtedness to Gregory Bateson and his patterned connectionism of systems thinking. More than any other, he helped me to understand the true nature of mind. Anthony Kronman, longtime professor at Yale, gave us all a timeless gift in his book, *Confessions of a Born-Again Pagan*. His book, which distills thousands of years of Western philosophy, clarified for me a new synthesis as a refreshing antidote to the received narrative that now has the world in an epistemological crisis. Beyond these, there are a whole host of influence. But read on, they are all in the book.

A beautiful October morning is closing out summer's books as I write. The land grows quiet. Trees will soon be bare of branch. The pond will turn to ice and the mountains will be covered with hoarfrost resplendent in the rising sun. Winter will blanket the land as bear and beaver huddle in their chambers. And then the cycle will repeat, again, and again. Earth—my greatest teacher.

Introduction

"Science advances one funeral at a time."
— Max Planck, paraphrased
"All our old stories are crumbling, and
no new story has so far emerged."
— Yuval N. Harari[1]

IN THE YEAR 1974, archeologists were digging in the Awash Valley of Ethiopia when they discovered the skeletal remains of an early female primate they named "Lucy." Fractures indicate that Lucy may have died falling from a tree. Although the evidence is inconclusive, Lucy could be the "Mitochondrial Eve"—the grandmother of the human race. She lived 3.2 million years ago.

Since then, her descendants have evolved, multiplied exponentially, and become the dominant force on earth, to the point that anthropologists now refer to the modern age as the Anthropocene— the age of man. 400,000 years ago, we came down out of the trees and harnessed fire. Study and comparison of the human genome indicates quite clearly that the forerunners of modern man migrated in waves out of Africa, beginning 200,000 years ago, with the famous Neanderthals eventually inhabiting Europe. Modern man migrated 100,000 to 70,000 years ago, again out of Africa, to breed with and outwit these forerunners. Today, many of us carry from 1% to 4% of the Neanderthal genome.

For the last 400 years, we have been on a tear: building, communicating, inventing, learning, and reproducing. In the process, we caused massive extinctions and depleted large sections of earth and its natural resources. We harnessed fire again to drive all sorts of machinery. We harnessed the genomes of plants and animals to increase our food supply and its diversity. We have gone to the moon, digitized information, and increased our knowledge in every field. We are now hovering in labs, exercising a God-like ability to rearrange the code of life. Cognitive and computer scientists are laboring intently to develop artificial neural networks, genetic algorithms, and quantum computers. As these three grow in power and complexity, they may lead to non-carbon Algo-Quantum Man, whose powers and abilities will transcend those of carbon-based man, signaling his obsolescence.

Here in the West, people enjoy a quality of life and abundance of food and material conveniences unheard of in former ages. No king in the year 1500 AD had the material luxuries of today's average Westerner. Who would have imagined that the common man could enjoy a hot shower every morning, a private auto to speed him down the highway at 60 mph, an MRI to assess his painful knee, and an artificial joint to replace it? Wine from Australia, sea bass from Chile, greens and pistachios from the Central Valley of California, water from Fiji, and an iPhone from China—these things are still not available to the majority of humans.

Everything is accelerating by the day. Moore's Law, while slowing, is alive and well in multiple fields. Computing power has grown exponentially, as has foreign trade. Trillions of dollars flash around the world electronically in seconds. Some science books written twenty years ago are now antiques. Carbon-powered autos, furnaces, and homes are becoming dinosaurs in the face of renewable energy, LEDs, network interfacing, insulation, and heat-pump technology. But at the same time, there are dystopian omens in the air. Everything has a price.

America is today the richest nation in the world yet ranks 19th on the happiness scale. Financial inequality is increasing rapidly, creating class envy and dissatisfaction. We are becoming more obese, with rising attendant problems of diabetes, heart disease, and cancer. Since 2009, America has had 57 times more school shootings than the other six of the G7 nations combined. Recent annual statistics indicate that 40,000 Americans died by shooting, 70,000 by drug overdoses, and 47,000 by suicide (50% by firearms), and there were another 1.4 million attempted suicides. Something is terribly amiss.

All political, economic, and religious institutions are built on stories—narratives of how the world functions and moves forward. Unfortunately, our main narratives—the stories that order our world and reality—have become threadbare and are now seriously wanting. Since our self-story is entwined with cultural and religious stories, it is no surprise that individuals and society are experiencing severe stress as the current story unravels, with no new and valid narrative to replace it. The stories for each of the following areas of society is buckling under internal failures and inconsistencies.

Politics: We have entered Orwellian territory where lies are truth and truths are lies. Lincoln's vision of a government of, by, and for the American people has ended up as a captive of the wealthy, multinational corporations, and special interests. On a multitude of issues, from a living wage to Medicare for all to global warming, the majority of our citizens are in favor of public policy that benefits the earth and future generations. Nevertheless, our politicians are more indebted to wealthy donors and bankrupt ideologies and consistently thwart the will of the people. Tax cuts are increasing inequity. Relaxed environmental regulations are not making America more competitive but are lowering the quality of the air we breathe and the water we drink. Although education improves a nation's competitiveness, both local and national budgets consistently give school funding low priority. Our political structure claims to respect national sovereignty, yet we meddle in the affairs of lesser nations for economic and military advantage. For politicians of a certain stripe, earth is America's resource colony.

Economics: Adam Smith, the economic philosopher and father of free-market capitalism (the concept on which the U.S. economic system is built), allowed that markets free of government interference would both regulate themselves by natural selection and lift society as a whole. He also believed that monies were to be kept at home in order that those profits could serve to enhance the structures of a free society and in so doing make that society even stronger and more competitive. Unfortunately, free market capitalism exists mainly in theory for large corporations, Wall Street, and the Big Banks. At every turn they seek government favors, protection from competition, bailouts, and subsidies at the expense of taxpayers. Why else would they need an army of lobbyists? As far as capital staying at home, it doesn't. All of the major corporations play accounting games to move cash to low tax countries, tax havens, and shell corporations designed to avoid paying taxes to the government of people who supported and made them successful. The main business model of Wall Street is to extract as much equity as possible while passing risk to the general public. By political appointments to the government departments intended to regulate them, corporations and financiers are, in effect, the foxes guarding the henhouse.

Religion: The pillars of religious institutions that once supported and gave meaning to the American story have begun to buckle like the columns of the Temple of Dagon thrust apart by Samson. With each passing week, states and dioceses release the names of pedophilic priests. And at a time when multitudes of refugees are fleeing and starving, the Vatican sits on inestimable wealth. Protestantism is in no better shape. Television evangelists have fallen like overripe apples from a tree due to financial and sexual scandals. Religious leaders supporting a corrupt regime are doing more to discredit the cause of Jesus than an army of atheists. Droves of people are rejecting this hypocrisy, especially the young, and finding their way out the back door. Finances are drying up and churches are closing by the thousands.

What can we make of these ideological anomalies in our political, economic, and religious institutions?

In 1962, historian of science Thomas Kuhn published *The Structure of Scientific Revolutions*. Critics place it among the top 100 most influential books of the 20th century. Although he was not the first to use the word *paradigm*, Kuhn brought the term to the attention of the scientific and philosophic communities. Aristotle used the word in his *Rhetoric* to teach his theory of argument by using examples. The word *exemplar* catches the meaning. A literal translation from the original

Greek is "to show beside." Essentially, a paradigm is a metaphor—a concrete example reflecting the pattern of an abstract concept. A paradigm is a mental picture that encodes a material reality. Religion, politics, economic theory, and cultural memes are ideas—immaterial concepts that serve as patterns for living and thinking. But since the brain thinks in pictures, we need a picture to help materialize the idea. Paradigms are the powerful metaphors by which we reason and live.

Kuhn sought to understand the nature and causes of scientific revolutions, that is, what it was that caused a sudden change in the ways we understand reality and the way the world works, what Kuhn called a paradigm shift. At the time, most people assumed that scientific progress was a linear process proceeding from one discovery to the next in a logical and timely series. What Kuhn discovered, however, was that science (and social institutions, as well) move forward in fits and starts. People accepted the geocentric concept of the universe for millennia, and then along came Copernicus. Newton was nudged aside by Heisenberg, Bohr, Planck, and Einstein. Darwin shut the gate to the Garden of Eden. And while these changes had gestated for some time, they erupted quickly to overturn existing orders. But what is it that precipitates these upheavals, these paradigm shifts?

In a word, paradigms crumble when they no longer provide a logical understanding of the way the world works. As paradigms once thought true are proven false, individuals and societies enter crisis territory and become unstable. That crisis, particularly in America today, is an epistemological crisis. Epistemology (from the Greek, meaning "to stand on") is the branch of philosophy dedicated to the process of knowing. To make an argument or explain a position, one needs firm ground to stand on; every tenet, every argument must have a rationally defensible support. For science, this is determined by the empirical process, and via tests for falsification. If one argues that the earth moves around the sun, then where is the evidence supporting this supposition? Where are statistics that support the effectiveness of supply-side economics in distributing wealth to more of the population? What data support the concept that human activity is the prime contributor to global warming? If the data, statistics, repeated experiments under controlled conditions and subject to peer review do not support a theory, then that theory has no epistemological basis. It is an unproven paradigm. That dog won't hunt.

What was long thought to be true has become demonstrably false. Since the mental story we call *our self* is a composite constructed from our personal experience coupled with the social and institutional stories our society is built on, forces that destabilize our society's story may also destabilize our self-story. When the threads of our story begin to unravel, our sense of who we are may also unravel. Individuals become insecure. Meaning evaporates. Depression follows, then fear that may manifest as anger.

The most common response is to strike out against the new ideologies and their proponents who are unsettling our world. Rather than accept reason and new findings, people resort to methods of identity-protective thinking, such as rejection of the facts, rationalization, and efforts to silence or eliminate expression of a new paradigm. This is the rich emotional ground from which Nazism emerged in the 1930s and a new nationalism is emerging today in the West. It is fertilized by vested interests whose wealth and power are threatened by new social, economic, energy, and scientific paradigms. It is mainly the carbon-fueled industrialists and ultraconservative religionists who are powering the denial of science and rejection of reason today.

New theories and new paradigms are often born of crises. New life seldom emerges all at once but must gestate for a period of time beneath the surface. Although new concepts that give rise to new paradigms may exist for long periods of time, they are often suppressed by the dominant story. As that story fragments, the gestating paradigm begins to make more sense to more people. As is often the case, however, those whose sense of self and reality are strongly bound to the dominant paradigm may never accept the new story. Hence, Planck's prescient quote that, "a new scientific truth does not triumph by convincing its opponents and making them see the light, but rather because its opponents eventually die, and a new generation grows up that is familiar with it."[2]

This is the story of Lot's wife, and the children of Israel who could not enter the Promised Land until that generation that had come up out of Egypt died. Is it not also the story of Nazi Germany, or creationist theology, or the Cartesians? But what of the new story?

I am suggesting a new story by which we can see life, the world we mutually inhabit, and the peculiarities of our own existence in a new light—a light that gives meaning to our separate lives and weaves us into the greater fabric of the universe in such a way as to yield a deeper sense of connection and belonging. The new story will contain fundamental elements of a story that has been with mankind for a hundred thousand years as well as new scientific truths of how life and the universe operate.

The great German sociologist Max Weber (1864-1920) popularized the term *disenchantment*. According to Weber, science, reason, and the bureaucratization of the social order had stripped the world of the magic and meaning previously provided by myth, art, ritual, and religion, leaving man to exist in a "polar night of icy darkness." Reason had killed the old God, industry was depleting the world, and World War I had slaughtered the nascent dreams of a new humanist order. It was a wrenching swing of the pendulum from the previous generation, the generation of William Wordsworth (1770-1850).

Writing his poem "Lines Above Tintern Abbey" in 1798, Wordsworth revels in enchantment: "And I have felt a presence that disturbs me with the joy of elevated thoughts; a sense sublime of something far more deeply interfused, whose dwelling is the light of setting suns, and the round ocean, and the living air." He would be followed by the romantic artists of the Hudson River and White Mountain schools, expressing their dreamy optimism of a new world unfolding amid the luminescence of wild nature, thundering cataracts, total freedom, and boundless potential.

Is it possible that we can accept quantum mechanics, evolution, the electrochemical nature of thought and intellect, the absence of a personal God and the divine right of human dominance, and still exalt in a world full of beauty, mystery, and harmonious order? Can we find new meaning and freedom from the fear of death and loss? Can we learn to belong again to something greater than any self and know that because we are part of that thing, we continue on with the whole? Can we accept that nothing is ever lost, but comes around eternally? Can we understand that belonging is both our birthright and the deepest meaning of love? Can we believe that all of time and the universe exists in every cell and particle of our bodies? Can we learn that there is no disjunction between us and others and the greater cosmos? The future of man, earth, and personal happiness hinges on our answers.

The following is my attempt to weave existing information together to form a novel tapestry of how the world came to be, how it is ordered, how it operates, the nature of God, mind, emotion, the universe, causality, and where we fit within the entire picture based on history, the sciences, and the humanities, and framed in a somewhat poetic point of view. **My goal is simple: A reconnected world is a re-divinized and re-enchanted world where we belong once again.**

1. Information

A bit of information is, "a difference that makes a difference."

— Gregory Bateson[1]

"Tomorrow, we will have learned to understand and express all of physics in the language of information."

— John Archibald Wheeler [2]

"What lies at the heart of every living thing is not a fire, not warm breath, not a 'spark of life.' It is information, words, instructions."

— Richard Dawkins[3]

Change and Flow

THE CALENDAR SAYS IT IS SPRING. The land says it is winter. A pair of geese returned this morning only to find their traditional summer home locked in ice. They stood for a time atop one of the beaver huts and looked about, anxious to get on with the business of raising a family ...

It's the end of March and I'm snowshoeing through a mixed forest, circumventing the vast wetland in front of my cabin. It was 14°F when I awoke this morning and the snow, which melted somewhat last week, is a hard crust that the snowshoes bite into with each crunching step. I will not sneak-up on any deer this afternoon.

The sky is clear blue. The snow reveals where deer crossed the path last week, as well as older depressions left by splay-footed snowshoe hares in the balsam thicket. Coyotes, foxes, and bobcats made soft impressions in last week's snow melt. A pile of splinters sits on the snow at the base of a rotting white birch where a pileated woodpecker has been searching for insects.

Some goldeneye ducks are in the open stream flowing into the bay, anxious for more open water. A friend remarked the other day that the ducks seem to have made a mistake by returning so early. I responded that it wasn't the temperature and open water so much as the increasing photoperiod that drives the birds to return. Their inner clocks tune to the angle of the sun, telling them that it's time to think about mates and nesting. Slightly swollen tree buds also sense the changes. Perhaps that same light is the source of the uplift I'm feeling as the days lengthen. Information is everywhere, seen and unseen, and it shapes both the thoughts and the actions of all living things.

Fundamental Reality of the Universe

For vast millennia of human history "believing was seeing." According to indigenous religions, the world is populated by innumerable and unseen spiritual forces that are the ultimate reality. Carl Sagan found a title for his book, *The Demon-Haunted World*, in The Isa Upanishad—"there are demon-haunted worlds, regions of utter darkness."[4] Blessings and curses, success and failure were in the hands of spiritual forces that could be placated by incense, offerings, and priestly intercession, or turned against individuals based on doubts, fears, or unbelief. Galileo and Newton stormed this stronghold by their appeals to the empirical process, mathematics, and hard data, and moved us from the *Age of*

Superstition to the *Age of Matter.*

Then the paradigm inverted to "seeing is believing." The "real" became what one can see, hear, smell, taste, or touch. The "non-real" was that which lacked substance, mass, physical presence. Reality had a material basis. Apart from material evidence, one cannot be convicted of a crime. Apart from physical and replicable material evidence, a hypothesis remains a hypothesis. But over 100 years ago, fractures began to show in materialism. Science was moving beyond *Materialism* to the *Age of Energy.*

Einstein's theories of special relativity (1905) and general relativity (1915) were the first hammers to fall. By equating mass with energy, Einstein demonstrated that there was another, and perhaps greater, reality behind material presence. In the case of nuclear fusion, it is not the hydrogen of the sun that runs the solar system, but matter converted to energy during fusion that is the causality behind winter and spring, photosynthesis and the pasta on our plates, the desire to cool off with a dip in the lake. Thought that is now driving my fingers on this keyboard is derived from differences in electrical potentials in the neurons of my brain that are energized by last night's beans and potatoes that derived from photosynthesis that was run by photons of light that were released by the conversion of matter to energy and radiated from sun to Earth.

An even bigger hammer fell in the form of quantum theory. Here, reality retreats even further into the quantum wave function, and is described using terms like entanglement, complementarity, superposition, and interference. The reality of gravity, matter, energy, and perhaps even time are reduced to immaterial fields and indeterminate wave functions. Although one cannot see electromagnetic, nuclear, or gravitational fields, our world runs on and is directed by them.

And if we continue the reductionism that has always been at the heart of scientific endeavor, all matter, all force, all of the quantum wave functions are reducible to information. The wave function carries all the information about an electron. Information is immaterial because the wave function is immaterial. Our five senses can only describe matter and energy which are effects of information. "Tomorrow..." wrote John Wheeler, one of the last collaborators with Bohr, Heisenberg, and Einstein, "...we will have learned to understand and express all of physics in the language of information."[5] Thus, began another paradigm shift: the *Age of Energy* to the *Age of Information*.

Information means what it says —"to form within." It is not unlike money that has no meaning in and of itself but is useful to transfer the value of one thing into the value of another. A farmer may harvest his wheat crop, take it to market where he transfers its value to money, then take that money to buy a new tractor. Thus, wheat is transmuted into a tractor via the magic of money. Money, derived from wheat that once grew in a field, has "in-formed" a tractor inside a barn. A house becomes a reality via the immaterial information of a blueprint. Shackleton's heroic adventure in Antarctica was reduced to information and formed again in my mind via the book *Endurance*. Because the code was logical, meaning and understanding were transferred. The writer communicated—established common understanding—with the reader.

Information can be transferred from one medium and in-form into another by logical patterns we describe as algorithms. Claude Shannon digitized this model according to Boolean algebra and introduced

Information Theory. Accordingly, all information can be digitized into 0 and 1 in an algorithm that serves to "in-form" thought as material pattern. Program coding was born—a digital language that could carry every bit of information about anything.

During this same period, the life sciences were making equally bold strides. In the mid-1950s, Watson and Crick described DNA as the carrier of hereditary code composed of phosphates, deoxyribose sugar, and four nitrogenous bases—adenine, cytosine, guanine, thymine (plus uracil that takes the place of thymine in RNA)—shaped as a double helix with the four bases bonded in pairs to make the rungs of a spiral ladder. Everything that can be known about your physical nature and hereditary past is found in your DNA. Information carried in the coding of DNA is causal for the formation of every protein, contains instructions to build every bodily system, makes brains, kidneys, toenails, tadpoles, and trees. And it functions in a nearly flawless manner with a remarkably small error rate. **We are informational structures through which matter flows.** Non-material information is converted to a physical structure when genetic codes from father and mother combine to form a zygote. Once formed, the zygote goes on to grow and differentiate into all the various tissues and organs that form a logical pattern we call a human body. Information is causal. How so?

The novel Coronavirus currently laying siege to human bodies and the global economy is essentially genetic information. It is information, contained in RNA code, that invades healthy human cells, neutralizes their defenses, then turns human DNA into a 3-D copy machine to churn out more viral malware that rips through our body, laying waste to systemic integrity and often ending in death. Everything about this cycle of destruction is due to false information. Likewise, the cycles of destruction that have preyed on human culture for millennia are based on false information contained in memes (units of cultural transmission), as opposed to genes (units of biological transmission).

Information is Relational

Anything of any importance in all of life is relational. Information is something relative to something else. If information is not related it is as meaningless as 00000000000. To say a batter hit the baseball 360 feet by itself is partial information. That number generally must stand in reference to the foul line, the outfield fence, other runners on base, and possibly a fielder's glove—the greater the relationships, the greater the meaning. Unrelated information is a nothing-burger.

Thus, there must be an "about-ness" to give information value. Information has meaning as patterned relationships. Pattern implies a web of connections that becomes a distinct unit carrying meaning. Logic is the reasonableness of fit of one piece of information with others. For instance, when I say elephant, you instantly pull up a mental pattern from past exposure. But what if I then show you the pattern of an animal with an elephant's trunk, the legs of a deer, the antlers of a moose, and the udder of a cow? The mind immediately rejects the illogic of that picture because it is a logical perversion. It is non-truth.

Most mornings, weather is the first app I tap on my iPad. It contains a whole bucketful of information—temperature, cloud cover, humidity, expected precipitation, wind speed and direction, and expected changes during the day. Depending on my plans for the day, all that information can yield a range of choices. Am I planning to spend the day in the cabin writing, or hike to the top of Mt. Shaw?

Shop, go out to dinner, or go kayaking in Lake Winnipesaukee? One day two years ago I kayaked the 23 miles of the lake, south to north. The most important piece of information in the forecast that day was wind speed and direction. Would the wind be at my back or in my face, and at what speed? A strong headwind would have scuttled the trip. Information reduces uncertainty and facilitates decision. The fundamental activity of life is processing information.

The Universe as Quantum Computer

Perhaps, as MIT physicist Seth Lloyd has suggested, the universe really is one grand quantum computer ranging across an ultimate reality of pure information, and with that vast trove of information cannot do other than experiment with random combinations to bring forth ever new and higher forms of complex relationships and exquisite beauty. "Every interaction between those pieces of the universe processes that information by altering those bits ... The history of the universe is, in effect, a huge and ongoing quantum computation. The universe is a quantum computer."[6] What new forms may yet appear?

According to Lloyd, life "is the grandmother of all information-processing revolutions."[7] Just as each succeeding generation stands on the accomplishments of former generations, algorithmic and organizational complexity rises additively with each new coded iteration. As a result, organisms grow in both complexity and in emergent qualities. The red eyespot of euglena becomes the eye of man. A fungal mycelium is the prototype for a neuron. The energy to make everything, from eyes to elephants, was derived from quantum fields, became matter, and was in-structured by information. But information is ever and always the primary driver—"…the universe is nothing but bits—or rather, nothing but qubits."[8]

Dualist and Linear, or Monist and Cyclical?

Imagine that you are living as part of a hunter-gatherer clan in a forest on the edge of the vast Serengeti Plain of Africa 70,000 years ago. Your kind is now the dominant species. Even lions avoid you. By day, men go off to hunt in groups and return with meat to share with the clan. Women stitch clothing from hides of slaughtered animals, cook food, care for children, communicate and facilitate relationships, pick nuts, berries, and fruits. Once darkness falls, people gather around a large fire to keep warm, review the clan's oral history, talk about the events of the day, and invent a new frame for understanding their place in the world.

Obviously, at this point in time, there is no scientific explanation for a multitude of phenomena: Why is there a rainy and dry season? Where does fire come from? Where does the sun go at night? How do other people get into my head at night and try to kill me (dreams)? Has another man cast a spell to make me sick? Did my child die because I killed the young of a zebra? Who makes all those points of light in the night sky? In a setting such as this, story was invented to make the world more intelligible—to provide information that connected man, the external world, and cause.

As millennia came and went, stories were amended, refined, codified, and translated into non-verbal media. I have seen them in ancient artifacts of stick figures, carved wood and stone, clay people and animals, petroglyphs on sandstone walls deep in hidden canyons, red ochred handprints and Kokopelli, demons and slithering serpents. They are all stories carrying messages about people and their attempts to understand the connections that existed between them and a world that was both seen and unseen. This striving to understand connections was the birth of religion. And it was all about information, both real and fantasized.

The words "story" and "myth" are synonyms. They are our best attempts to navigate the shadowlands between the known and unknown. The goal of story is to provide information to help us understand our place relative to time and the world. Fantasy and metaphor—both figurative speech—are not intended to deceive but to give form to ideas that transcend literalism. Perhaps this is why multitudes, including Einstein, wrestled mightily with the paradoxical and seeming contradictory world of quantum mechanics, thinking it more closely aligned with metaphysics (the world of ghosts, golems, and God) than physics.

Therefore, as you sit around that communal fire 70,000 years ago, at night, on the edge of the Serengeti, listening to stories told by elders, you slip out of the reality of the light world and enter the alternate reality of a dark world. In your light world, things were reasonably straight forward, guided by five reasonably trustworthy senses that gave you most of the knowledge you needed to navigate life, Earth, and community. The light world was knowable, interconnected, unified, and reasonably understandable. But here in the dark world there are unseen spirits, often malevolent, that are powerful and could do great harm. They could be placated with prayers, chants, sacrifices, and offerings. But one needed a specialist in such matters—one needed a priest, a shaman, a holy man to prescribe and guide offerings and sacrifices. Welcome to the roots of institutional religion. Welcome to the world of dualism.

Thus, for tens of thousands of years, people lived in two different worlds. The light world was relational and reasonably understandable. Reciprocity ruled. Rains came, plants grew and blossomed, people ate. We were dependents of Earth. As was Earth, so was man. Eventually, we learned to facilitate fruition by planting a few seeds in plots that we could come back to as supplements to hunting and gathering. We learned that the best plots were in river deltas, with their higher fertility from yearly silting and moist soils. With success, plots gradually grew bigger and people remained longer in one place.

With more food calories, especially from cultivated grains, populations grew larger in these fertile river deltas. As population grew, the demand for labor grew. With more people, specialization increased. The more intelligent became managerial, and the less intelligent were consigned to labor. This was the beginning of the first city-states of Mesopotamia. And it was the start of the struggles and destructions that are tearing at our world today.

As society became urbanized, the felt connections with Earth and the natural order began to fade amid a life of labor in one fixed place. Life was now in the city and equated with order, light, and the forces of goodness. Meanwhile, the world of nature was consigned to darkness and the haunt of evil that would undo what light had built. The people who continued to live out there, and from time to time raided the city's grain stores, were savages, pagans, barbarians in service to the arch-demon, Lucifer, Satan, the Devil. They were the Sons of Darkness who sought to crush the Sons of Light. False assumptions began to pile up.

As populations grew, technologies increased, societies became more stratified in castes. Some governed. Some were craftsmen. Some managed and others became priests and scholars. But the greatest number were laborers and farmers without whom nobody would eat. And people being people, this set up a visible inequality that fostered envy and conflict. Manipulating and managing people to prevent rebellion and escape became the job of politics and religion.

Information became the management tool of choice. The elite soon learned that those who control the story, control the information, that controls minds, that controls bodies, that creates profits by which the elite can live off the labor of others.

Thus, information provided "mind-formed shackles" (Wm. Blake) to keep people fearful, pliable, subservient, and willing to pay taxes for protection against a mostly conjured evil. Little has changed down to the present. This is the world of religious dualism.

In dualist understanding, the world is divided into antagonistic camps—the forces of light and dark, good and evil, God and the devil. Life is a perpetual conflict where soul-destroying evil must be exposed, resisted, and oppressed at every turn. Earth also becomes suspect as an agent for enticing man's allegiance away from the all-powerful but unseen God who resides in heaven far away. Matter is subservient; spirit is dominant. In this model, obedience to (male) spiritual authority is the pathway of salvation. Females cannot hold spiritual authority because of their role in the mythical fall of man and menstruation that makes them unclean (Leviticus 15).

The world of dualism is linear. It began with the divine fiat of God at creation and will achieve its glory in a final apocalyptic battle between light and darkness in which the forces of light and God will be victorious and usher in a time of cosmic peace and harmony. This is the basic linear paradigm of Zoroastrianism, Gnosticism, Manichaeism, Judaism, Islam, and Christianity.

Perhaps these dualistic religions—mere variants of one another—survived and morphed into more sophisticated forms because they provided a highly effective paradigm for controlling people without the need for proof (one cannot demonstrate an immaterial God, heaven, the coming judgement, resurrection, or eternal damnation). Religion is able to captivate and control people with a received narrative. A religion that does not support and benefit the governing elite will never survive.

Monism, on the other hand, is derived from the natural order of things. Therefore, Monism and Naturalism are nearly synonymous, each favoring a more ecological narrative of how the world is ordered and works. All things flow together and are a part of each other. Just as one season derives from the former and flows into the later, all life and matter are intertwined in one grand tapestry, where each part is related to and affects the others. Perhaps the monist paradigm is best expressed in the *Gaia Hypothesis* of Lynn Margulis and James Lovelock. In this view, Earth (Greek god = *Gaia*) is one unified system, with inseparable flows of information between a multitude of microsystems that talk to each other for balanced continuance (homeostasis) of the emergent metasystem (Gaia). Control and command are distributed within the system, without the need for elite control or management. Monist systems operate by distributed feedback because the survival of the whole is the supreme value.

Monism is ascendant today because both science and casual observation lend abundant support to the monist narrative of the way things work. Geology demonstrates a clearly uniformitarian process in the evolution of rocks, minerals, and major landforms. Modern genetics can show no clear disjunction between the forms of life on Earth today. Our genes—frogs, fish, fowl, and man—are edited and recombined versions of one another. Similar genes command similar functions across broad ranges of living things, that, though they appear dissimilar on the surface, function in nearly identical ways based on nearly identical gene/protein/enzyme determinants.

As I write, a beaver is sitting across the marsh chewing on an alder branch. His mitochondria will turn the derived nutrients into energy in much the same manner as mine with last night's pizza. As a result of time spent in the Sierra wilderness, John Muir noted that upon close examination one would find "…everything hitched to every other thing". Separation is a false narrative.

These arguments are critical both to our understanding of the world and how it works, and to the transition to a new paradigm that exposes the anomalies in our current stories and enables us to move from today's crises into a narrative that brings us into a more peaceful, less harmful, and sustainable future. Let me explain…

If information "in-structures" itself within us, then it is imperative that the structure be trustworthy, substantial, capable of bearing the loads of life. False and insubstantial structures are forever and always hurts looking to happen. If information is causal, then our lives and actions cannot affect goodness and peace if the informational basis is corrupted or false.

Now, taking this a step farther, consider all those religious texts that adherents claim to be the "Word of God." If a male Jew wrote that text, would it be a surprise if he claimed that male Jews were God's people chosen to rule this world? Or a Baptist, Catholic, Mormon, Jehovah's

Witness? Claims to religious authority, divine revelation, membership in the "chosen few," or special access to divinity are merely a tribal ruse to protect a sense of entitlement. For the myth of a loving God who made them superior, some are willing to crush the light from the eyes of untold billions who are also the image of God. But wouldn't that make them agents of the devil?

Observations: Quality of information matters. We seldom get good outcomes from bad information. If information cannot bear scrutiny it should be suspect. Although they sometimes make us uncomfortable, facts are our best friends because friends tell us the truth. Two questions of utmost importance: Where did you get that information, and how do you know it is true?

The Informational Field Hypothesis and Paradigm Shifts

This sylvan cabin, where I sit and think about time and ideas, was constructed to be the best seat in a theatre where the "Nature Channel" plays 24/7. No two moments are ever the same. On blue sky October days, migratory birds land and take off from the pond. Clouds rub their bellies on mountain peaks. One season blends into the next. Leaves paint the land in muted shades and vibrant color, then return to Earth. Crows caucus; marsh hawks hover; eagles swoop. Mink and snowshoe hares write their stories across the snowfields of January. And now that mating season is in the offing, geese and songbirds are clamoring for mates and nesting sites. Nights will soon be filled with "hyla bells"— the songs of ten thousand spring peepers peeling out a longing for life. Thus, I oriented the cabin to the marsh, pond, mountains, and open sky because that's where life, energy, and variety are greatest.

Greater concentration and variety are a density of information that brings novel orderings. Increased information generates a force field that both attracts and facilitates even more information in multiple complex emergent forms.

"In physics, a field is a region in which each point is affected by a force"(Britannica). Modern physics began with Newton's formulation of the gravitational field. Faraday and Maxwell later explained the electromagnetic field. Then came Planck, Bohr, Heisenberg, and Pauli to propose quantum field theory. Modern physics explains matter and energy on the basis of Field Theory. We now live in what is called *The Information Age* that originated with Claude Shannon's Theory of Information developed at MIT and Bell Labs. Others have taken the concept of information further. John Archibald Wheeler is famous for his pithy *It from Bit* view that everything is derived from information. Taking all of this into consideration, this marsh and forest and mountain are suggesting one grand synthesis. May I hypothesize?

Information is the basic reality of the universe, and the above-mentioned fields are patterned manifestations expressing states of information. In other words, "Informational Field" (a force field that attracts even more information) and "Universe" are synonymous—matter, energy, and the fields they generate are derivatives and expressions of information. Matter and energy are states of information, as steam, ice, and liquid are states of H_2O. Sexual attraction is the potential of achieving new informational combinations to cope with changing environmental parameters and send that information into the future. Consciousness is a node of concentrated information that is aware of itself. If information is fundamental to the universe, then consciousness is an attribute of information and not necessarily carrier dependent. The greater the density of information, the stronger the attraction of the informational field surrounding it, just like gravity and electromagnetism. Increasingly concentrated information may soon evolve to meta-carbon life with levels of consciousness and logic that transcend the narcissistic egoism currently trashing Earth.

New Information Requires New Forms

In every age, as a new ordering of information and understanding become widely accepted the old forms pass away. This is an absolute necessity because new information is dynamic, moving reality from old form to newer form. The parable of the wineskins is an apt metaphor...

"Neither do men pour new wine into old wineskins. If they do, the skins will burst, the wine will run out and the wineskins will be ruined." — Jesus of Nazareth

Two thousand years ago, wine was stored in "skins," generally made of goat hide. New skins were fresh and supple and could expand and contract as the new wine continued to ferment and mature. With age and the stiffening of the hide from alcohol, however, the skins became rigid and less yielding. To be economical, winemakers would attempt to reuse old wineskins for new wine. Perhaps it worked sometimes, but more often it led to disaster. As the new wine fermented, it stretched the old skins beyond their limits. Unable to expand, the skins blew out, ruining them for future use and spilling the good new wine. All was lost.

The obvious point is that new ideas are dynamic and expansive, requiring flexibility and a willingness to adapt. Vested interests most often refuse to change because change means loss of security and power. Can the Vatican all of a sudden, in response to new revelations of sexual predation, declare that celibacy is an unnatural human dictate and accept married priests? Will Exxon/Mobil agree next week that fossil fuels are devastating the Earth, declare their reserves stranded assets, and move into wind and solar power? Can conservative politicians accept that health care is a basic human right and support Medicare for all without losing significant political capital? In each case, vested interests appear so ossified that they will fight on to the point of rupture. Progress often does come one funeral at a time ...

Even by New Hampshire standards, this past winter was long, cold, and snowy. As a result, nearly 18 inches of dense, crusty snow-ice had accumulated on the metal roof of this cabin. It's been stuck in place all winter, only melting in drips as the weather warmed this week. Yesterday was in the 40s and remained in the 30s overnight. We had a good rain two hours ago. The rain ran under the ice and lubricated it against the metal. Then, just ten minutes ago, there was a roar and the cabin actually shook as an avalanche of ice slid off the roof with the sound of rolling thunder. The final stage of a paradigm shift often comes with the same speed and intensity.

Paradigm shifts occur when a greater number of people understand new information. Things and events have causes. But what causes information to synthesize into monkeys and men? What is the **ultimate cause** behind the multitude of patterned informational forms we see every day? What is the nature of God?

2. God Is Meta-Logic

"By the word of the LORD were the heavens made, their starry host by the breath of his mouth. He gathers the waters of the sea into jars; he puts the deep into storehouses…For he spoke, and it came to be; he commanded, and it stood firm."

— Psalm 33 selected

"In the beginning was the Logos, and the Logos was with God, and the Logos was God…Through him all things were made; without him nothing was made that has been made. In him was life, and that life was the light of men."

— John 1 selected

"All great truths begin as blasphemies."

— George Bernard Shaw

Personal Journey

IT'S SUNDAY MORNING and my father is wearing his only suit. Mother is in one of her better dresses, shielded by an apron, as she slices onions, carrots, and potatoes to go in the oven with a pot roast that will cook slowly while we are in church. I am perhaps four years old. Upon arrival at the Brownsville Baptist Church, mother takes me to the Sunday School preschool class while she and my father join in the adult class. In an hour she retrieves me and takes me to be with them in the worship service ...

For the next 60-years, my life revolved around religion, eventually leading me into the pastoral ministry. For two decades, after receiving an MA in Theology at Wheaton College Graduate School (epicenter of evangelicalism), I served as pastor for local church congregations.

During that time, I watched as evangelical Christianity evolved into a quasi-political movement, with the Moral Majority, 700 Club, Christian Family Association, etc. In the process, national Christian leaders and spokespersons aligned ever more closely with political leaders and public policies that are a complete repudiation of the life and teachings of Jesus of Nazareth. More recently, 82% of white evangelicals voted for the presidency of Donald Trump. At that point I felt a total disconnect.

In the process, I reevaluated the stories my life was built on, threw out a lot of garbage, and accepted new patterns of reality. I embraced a new story of life, God, the universe and our place in it.

Logic

What is God? Where is God? Is God a person, and if so, what is the form of that person? Will we actually see God someday, perhaps floating on a cloud? What would it be like to be continually in the presence of God? Where does God live? If God is not a person of carbon form, what is the substance of his being? Or, if God is pure mind, does he need a form? Or could God appear in a multitude of forms? What part of us can be singled out as "the image and likeness of God"? How does an immaterial God manifest as material form?

These are children's questions. We've hemmed and hawed over

them for centuries, but have offered few, if any, substantive answers. Metaphor is the best we can do, like Michelangelo and William Blake did through art and poetry. It's hard to shoehorn ever present, all powerful, all creative, all knowing immateriality into a brain that likes pictures and patterns. Many have tried the verbal/philosophical approach. Few have done better than the ancient Greeks.

As the Hellenic pantheon of Homer's time began to buckle under perceived anomalies, Greek society started to fracture. City states had become vulnerable to opposing forces. The old stories were losing credibility in the face of a nascent science and cultural cross fertilization. If the old gods were mere fictions—deities cast into faulty human likenesses with all of our jealousies, intrigues, and fickleness—what understandings would rise from the ashes to address the new realities of life in a world of atomic theory, religious foment, political flux, and global trade? It was a period not entirely unlike the 21st century.

Central to this new concept of a dematerialized God was the word *LOGOS*. Logos is translated simply as "word," as in the John 1 quote. Most people gloss over that term unaware of the richness it carried in Greek philosophic understanding. Readers of the New Testament may read the logos passage in John 1 and say, "OK, Jesus is kind of like the summary of the word of God in the Old Testament." They conflate Jesus with the verbal volition of God in the Genesis creation story or the revelatory word of the prophets. This is superficial compared with the depth and breadth of meaning imputed to Logos by the early Greeks.

The root of *logos* is *lego* = "to speak, to tell, word, and speech." It carries with it the sense of gathering, collecting, or organizing in logical patterns or order. Logos as logic demands to know *on what ground* a postulate stands. Correct logic provides a foundation that can withstand intense scrutiny and retain integrity. It is entangled and coterminous with "mind," "reason," "thinking," "truth," "law," "life," and "nature." Interestingly, the word "nature" is the Greek *phusis*, from which we get our word "physics." Logos rises to the stature of a grand cosmological principle which is closest to Aristotle's understanding of God and connects man to both the world and deity. Logic is the interconnection that holds all things together. Truth is obtained by the logical interpretation of phenomena. This harmonizes with Aratus' poem *Phaenomena*—"In him we live and move and have our being," suggesting that God is the binding principle that both organizes and empowers all of life and the world. Logos is the fundamental order of the world. For Zeno, logos is *logos spermatikos*—the seed of reason that is born as both life and truth. For Epictetus, the man who has orthodox logic will order his life by following the ways of nature. This differs little from a life of following the Tao, as endorsed by Lao Tsu. Diogenes, another stoic, saw logos as the merging of the rational power of order with the vital power of conception. This is almost identical to the John 1 quote.

In Neoplatonism, logos is the molding power that gives form to living things and is closely related to idea (thought or concept), form, light, and life. The clearest representation of this merging is the early Greek Orthodox cross using *phos* (light) for the vertical beam, and *zoe* (life) for the horizontal beam. Logos manifests itself as light and life bound in an all-encompassing unity covering the four directional points. In fact, the opening of John's Gospel is a clear echo of Plotinus, "In the beginning, therefore, was the word and the word was all things…their words are harmony which unite the whole into one."

Plurality becomes harmonious unity under the order of logic.

Thus in Hellenistic mysticism logos is essentially a cosmic and creative potency, the guide and agent of knowledge, increasingly represented as a religious doctrine of salvation, the revealer of what is hidden…logos releases creative and constructive forces, and then takes them back into itself in an eternal process … in a gradual unfolding of being.[1]

In nearly every case where logos is used in ancient Greek philosophical and religious texts, one could substitute the word "God" without any loss of meaning. God and logos are coterminous. And the one word that summarizes most clearly the values and attributes of logos is its derivative—Logic. **God is Logic—rationality, reason, nature, truth, pure and integrated mind.** But standard logic cannot describe the multifaceted God of Nature.

Standard logic too easily ends in circular contradictions. It is rigid, inflexible, and sometimes gets tied in knots by literality. Think of a Rubik's Cube. There is one, and only one way to complete the puzzle, and that is to have all the same colors unified on each side—all reds, all whites, all greens, etc. That's the simple logic of the thing. But what if we spin parts through multiple axes and jumble the colors. You now say it is illogical, incomplete, unfinished, contradictory. Our daily logic is pretty much like this, cut and dried.

But just imagine that each block had the potential to simultaneously be every other color, depending on what the observer desired? Every red could also be green, white, blue, yellow, or orange, and vice versa. Now we have a Quantum Rubik's Cube with near infinite potential to be a gazillion different probable combinations. And…be all of those at the same time in a spinning cloud until we selected one probability by observation. Confusing, no? I cannot fit this into a little brain that runs mostly on a linear logic. I'm befuddled. Yet, this is a crude illustration of what quantum mechanics tells us of the fundamental level of reality.

This is why an omniscient, omnipotent, omnipresent Deity, who can be all things, know all things, and contain all power at once, cannot be contained by linear human logic. We need a *Meta-Logic*, or *Quantum Logic*.

Quantum Logic

Before you throw up your hands, the ideas around Quantum Mechanics (QM), quanta of energy, quantum field, etc. are hard to slice, even for the sharpest knives in the drawer. Richard Feynman, once described as "the smartest man in America" in a national magazine, is alleged to have said, "Any person who claims to understand Quantum Mechanics is full of baloney." I don't grasp a lot of what follows. These are the results of my research and conclusions of some of the smartest people on the planet. And the thing is, most of them say Quantum Mechanics is true because it has been verified over and over in a multitude of controlled experiments. QM is the most accurate way to describe the underlying realities of nature. Yet, even Einstein initially thought it was a weird fantasy that would eventually breakdown. But it didn't. QM is the foundation of modern physics. Now, onto our Logic quest.

The concept of Quantum Logic (QL) was born in a 1936 paper by George Birkhoff and John von Neumann, *The Logic of Quantum Mechanics.*[2] Their proposition was that QL is an ortho-latticed, multidimensional, many valued logic that surpasses the traditional yes-no of classical logic and Boolean algebra and is more descriptive of the quantum wave function that in superposition contains conflicting informational states which are simultaneous truth (like our Quantum Rubik's Cube). It is a meta-logic which is context/observation sensitive and yields answers based on differing contexts which are equally true and false, and holds all potentials, both past and future, as present reality. A photon is **both a wave and particle,** has momentum but no mass.

In classical (linear) logic, cause and effect often generate unreasonable paradoxes. Consider the Epimenides Paradox: Epimenides was a Cretan who said, "Cretans always lie." The larger quote is a class that becomes subordinate to the subclass of the smaller quote, making it into a paradox that creates an endless loop—if yes, then no, if no, then yes, ad infinitum—that would cycle until a computer stops. And yet, the universe appears quite comfortable with all sorts of seeming paradoxes. Religion has always embraced paradox. Science is just catching up.

Shiva is the Hindu dancer whose dance is both creation and destruction. Abraxas is the Gnostic god of both horror and beauty, day and night. Paradoxically, Jesus' wrenching death by crucifixion becomes the portal to immortality. Lao Tsu wrote, "Under heaven all can see beauty as beauty only because there is ugliness. All can know good as good only because there is evil." Chaos theory views bifurcation and collapse as the source of genesis. The Taijutsu (Yin/Yang symbol) contains black and white as equal portions of Fibonacci curves within a circle as the contradictory elements necessary for wholeness and balance. The Genesis account of creation states that light and being derived from a dark void—order born of chaos. According to the Internet Encyclopedia of Philosophy, "One might characterize Quantum

Logic ... in terms of a strategic subversive attitude towards classical logic and the very foundations of metaphysical understanding."[3] That means, it appears irrational. Dualism has it backwards—darkness does not destroy but gives birth to light.

Each of the above belongs to the domain of quantum, meta-logic, or meta, meta-logic that makes impossible demands on the limited logic processing available to the human brain.

And yet, if the human brain is a derivative of Quantum Logic (as it must be), then it seems reasonable to expect that its structure would reflect that logic, just as a painting reflects the mind of the painter. The human brain bears some resemblance to Birkhoff and von Neumann's multi-dimensional, multi-connected lattice. The neocortex does, in fact, possess a grid pattern of stacked lattices with pathways running in three perpendicular directions like a cube and similar to the crossbar switching of circuit boards.[4] Within this multi-dimensional lattice there are roughly 10^{15} connections.[5]

In *Physics and Philosophy* (1958), Werner Heisenberg (one of the founders of QM) said the probability wave was a quantitative version of the old concept of "potentia" (potential) in Aristotle's philosophy, "something standing between the idea of an event and the actual event...between possibility and reality"[6] In a wildly geeky daydream, I wonder (as suggested by Birkhoff and von Neumann), if we were to project the wave functions of every particle in the universe onto a lattice of infinite connections with every contextual event as multiple images transformed into an informational map of all time and space, could we not identify every probable thing, event, interaction, and result that might occur past, present, and future as a present reality? Potentials are time (which can flow forward and backward in quantum equations) and context/decision dependent realities that must manifest somewhere in a lattice of infinite dimensions and meta-causal information that appears paradoxical in isolation but holistically integrative in total.

This fits with Seth Lloyd's concept of the universe as a quantum computer, John Archibald Wheeler's *It from Bit*, and Holger Lyre's statement: "Any quantum object may further be decomposed or embedded into the tensor products of two objects, nowadays called quantum bits of qubits."[7]

Perhaps this isn't so wild. Various schools of String Theory propose a universe that contains between 10 and 26 dimensions, as opposed to our mundane three-dimensional world (four with time). How are we to wrap our limited brains around this? Things that are paradoxical/antithetical/nonsequiturs in classical logic might be seen as meta-logical holistic truths from a universal perspective. Isn't the resolution of antitheses the transcendent beauty of music that lifts us to higher dimensions? I believe the human brain knows things—deep things—that it cannot fully access consciously.

Yet, we sometimes get a fleeting glimpse into levels of reality not seen in common day. In a later chapter I will discuss transcendent states induced by near death experience, psychedelic drugs, and deep religious experience. Each case might come under the heading of "Unitive Consciousness," which I would define as a multi-systemic coherence where each system's wave functions are in phase and result in a constructive interference which manifests as emergent unities previously unimagined. Unitive consciousness reveals a merging—a trans-systemic identification that tends toward monism—of conscious mind with plants, clouds, rocks, and even deity. Observation is ever and always just one actuality (Aristotle's *potentia*) of nearly infinite possibilities.

Agencies of God

A God of Meta-Logic meets part of the standard definition of God, i.e. that God is immaterial, unseen, dwells in light inaccessible, has no physical needs or dependencies, is pure being ("I am that I am"), cannot be contained or diminished, is omnipresent, omnipotent, and omniscient. But immateriality or pure conception needs agency. Apart from some form of agency, logic is impotent.

For instance, I conceived this cabin that functions as a place to write these words. That concept, however, would never have taken form apart from hands and legs and eyes and a back, plus the tools employed by them. Logic or reason was the prime mover, but the prime mover needed something to move through before creation could occur. Information must be substantiated into matter and energy before it can emerge as a logical informational pattern. Thus, a person seeing this cabin for the first time would have to do some reverse engineering to understand how it came to be. He would see tools in the tool shed that obviously found their purpose in the shaping of boards, cement, nails, windows, and doors into the logical pattern that is the cabin. The pattern is logical because it is unified and harmonious in form and function according to the original concept. But the tools are useless without hands and back to hoist and apply them in a multitude of diverse functions. Seeing my hands and body, the visitor would automatically make the connection of agency with tools (sub-agency). But he would never see the logic of conception unless I left a metaphor in the form of a blueprint. Thus, the fundamental necessity and prime mover in the entire process can never be observed or fully known to an entity outside itself. Observers can only see effects and reason backwards. God (meta-logic/mind) is the reason behind the reason of agency.

Logic is pure being, pure conception, pure mind. Logical informational patterns in material form are derivative being—phenomena. Force—a push or pull resulting from interaction with another object—is primary agency; energy is secondary agency. Force is defined by field. Energy is defined by potential. A magnetic field applies the energy of electrons in a stator to spin the rotor of an electric motor to move an object and accomplish work. Intense gravity in the sun fuses hydrogen into helium, converting matter to energy released as photons which energize electrons in chlorophyll to do the work of synthesizing carbon, hydrogen, and oxygen into glucose. The builder of cabins ingested that glucose in food to release the energy of photons from the sun that was stored chemically by chlorophyll to do the physical work of hammering nails. New informational manifestations derive from energy, which itself is stored information. Energy is transferred back to information via work.

Follow all these trails backward from work to energy to fields to forces and one ends at mind, at Logic. The agents of Logic, therefore, are the four primary agencies of the universe—gravitational, electromagnetic, strong, and weak nuclear forces. Manipulated by Logic, these 4-forces created everything that has been created. They are the tools of God. Logic is the mind of God, and God is pure mind. What Logic creates, therefore, can only be logical. Illogical and irrational patterns cannot continue in a universe governed by Logic. They are conflicted patterns, products of chaos and randomness, that are unselected by reason.

Illogical patterns are an unlawful mathematical absurdity—systemic insanities.

Let's combine all of the above and see how closely these postulates come to traditional definitions of God. Theologians have always defined God by attributes. Being human, we therefore cast God's nature and attributes in metaphors congruent with human patterns of information—we describe God in human terms (just like the ancient Greeks). What, then, are the generalities theologians have used for millennia to describe God, and do they fit with the above definition of God as Meta-Logic using the four forces of the universe as agents to in-form Logic across the universe as information, matter, and energy organized in logical patterns?

By review, the common catechism, from Augustine to Aquinas to Luther, understands God as omnipotent, omniscient, omnipresent. Biblical texts variously describe God as light, life, truth, power, love, just, righteous, eternal, immortal, invisible, unchanging, without form yet forming all things. God creates, sustains, causes growth, and is wisdom. God is without substance yet brings all substance into being. God is perfectly and completely actualized potential. There is nothing left to do that God has not or cannot do.

All of these attributes apply to Meta-Logic manifest through the four forces as light, life, truth, power, creativity, immateriality, pure being, and changeless continuance not contingent on any externality such as being, force, time, or condition. Without substance, it gives substance to all. Therefore, all things across all of time are extensions of God, who is pure reason (Logos). Physics, philosophy, mathematics, and observation validate Spinoza's God.

Baruch de Spinoza (1632-1677) was a Jewish-Dutch lens grinder, born in Amsterdam of Portuguese Sephardic parents. He laid the groundwork for the Enlightenment and modern Biblical criticism and is considered one of the greatest philosophers of the 17th century. Because of his controversial ideas concerning the authenticity of Hebrew Scripture and the nature of God, he was shunned at age 23 by the Amsterdam synagogue. The Catholic Church later included his writings on its Index of Forbidden Books. On the other hand, GWF Hegel, dean of German philosophers, said, "You are either a Spinozist or not a philosopher at all."

Spinoza has been variously branded as an atheist, a pantheist, and a heretic. His most well-known pronouncement, "Deus, sive Natura" (God, or Nature) equates God with the natural order of things, viewing the creator as being so closely intertwined with creation that one morphs seamlessly into the other. There is no separation in Spinoza's monist world, in contrast with Descartes' dualism that separates mind and body, God and nature, heaven and Earth. Spinoza is surrounded by God at every turn. Descartes must wander like an orphan, abandoned in an unintelligible and sinful world. The major religious systems of the West are all Cartesian.

For Spinoza, any one of the multi-variant manifestations of information, matter, or energy within the universe is an extension of the mind of God. And since God is pure Logic (rationality, mind) all things are God in-mattered, or incarnate. We can know God by studying what God does, just as a person can understand a lot about Dan Lake by observing what Dan Lake does. Look at this cabin and you get a large glimpse into the logic (mind) of its builder. If the essence of a person is thought and idea, not body, then this cabin is my thought and ideas—it is an extension of me. John Ruskin said much the same thing, "Tell me what you like, and I'll tell you what you are" (*Unto This Last*).

According to this model, dualism is an absurdity. There is no separation, no disjunction, between the mind of the universe and its multi-variant expressions. Golden leaves fluttering earthward in autumn, the massive white pine behind the cabin, the great blue heron on the edge of the pond patiently waiting for a fish—all are gateways to understanding the mind of God, living metaphors of the Logic that governs the universe. They are one and a unity. Mind divorced from matter is unintelligible. Therefore, God, the universe, information, matter, energy, and all of their manifestations are one, just as the multiple facets of a diamond are singular points of view that comprise one unified whole to reveal the logic of its crystalline structure. Spinoza was correct—the universe is Monist. The mind of God is within all things. Immaterial Logic is the basis of all that is. **Logic is the ultimate causation of the universe. Therefore, God is Meta-Logic, Meta-Mind.**

I submit that the infinite number of possible combinations of information, matter, and energy is a function of the logical interactions of the four forces with current environmental parameters. Rocks, mountains, water, oceans, life forms, and stars are all the result of logically patterned information generated by harmonious interaction of force fields and environment. I believe the same can be said of economic, political, and religious systems, as well as the rules of social and cultural engagement. Each must be the most logically efficient answer to the questions of the environment. Geologists understand that the same elements crystalize into different forms depending on the cooling environment. In biology and business this is called natural selection. The process is adaptation. The multitude of processes that, on the surface, appear chaotic, are in reality governed by a strict logic that calculates novel outcomes. Logic is the conductor in the orchestra of creation. Musicians are the agent/forces. Music is the product.

Earth systems, from oceans to atmosphere, are now flashing red telling us that we are acting illogically. These systems evolved across eons of time, adding, growing, self-correcting, complexifying, for sustainably balanced interaction with all other systems. Excess carbon dioxide, rising oceanic pH, air and water pollution, plastic islands in the ocean, and concentrations of humans, sealed off from nature, whose faces are more and more focused on electronic screens, are all illogical inputs that are destabilizing the systems that support life.

If God is Logic, then illogic is the atheism that has infected humanity. The illogic of our stories and actions is an attack on Logic. And that's a dangerous place to be, for a resistance of Logic is the ticket to a place of great hurt. The counsel of Logic is wisdom.

To summarize: God is the Meta-Logic of the universe acting through the agencies of the four forces to in-structure Logic's Being into logical informational patterns. There is no material or personal God in some heaven far away. God is immanent in the Logic of nature's patterns that sustainably continue within the grand scheme of the universe. Man is derived from this same Logic and, therefore, is a manifestation of divine mind. If there is a devil, it is the illogic of conscious, linear human purpose that destroys nature's (God's) logical patterns.

We need a bigger God, a God as big as a universe that is simply the manifestation of all that this God is. But we also need a God who is within us and with us at every step. This is the Logic that made us and is with us at every moment. We can never be where God is not, because we are a form of that God. If God cannot die, neither then can we. It is time for us to come home again, to the only home and only God there could be ...

Today's work is wrapped up. I've been thinking hard, wrestling with ideas that often feel out of reach. As usual, I turn my gaze back to the natural systems outside the cabin—forest, pond, wetlands, and mountains. Timeless Logic permeates them. Because they are logical, they go along, year in and year out, adapting, changing ever so slightly, complexifying with life. The ducks and geese are still walking on the ice, waiting on open water for food and nesting materials. They've returned based on a timetable honed into their genes and brains over thousands of years, and they will succeed in raising another generation. They evolved to adapt. When tempted to think I know more and better, I ponder this exceptionally rare and beautiful scene and remember this—this I remember—that it is all self-sustaining. Nature's Logic is superior to mine. It does not need me for continuance. The best thing I can do is keep my hands off. I don't want to become an atheist, setting my short-sighted pseudo-logic above eternal Meta-Logic, which is Nature, which is God in-mattered.

3. Logical Informational Patterns

"The great book of Nature can be read only by those who know the language in which it was written, and this language is mathematics."
— Galileo

"My central thesis can now be approached in words. The pattern which connects is a metapattern. It is a pattern of patterns. It is that metapattern which defines the vast generalization that, indeed, it is patterns which connect."
— Gregory Bateson[1]

"Now, as nothing is necessarily true save only by Divine decree, it is plain that the universal laws of nature are decrees of God following from the necessity and perfection of the Divine nature ... nature, therefore, always observes laws and rules which involves eternal necessity and truth, although they may not all be known to us, and therefore she keeps a fixed and immutable order."
— Spinoza

Consider the Eye

AFTER A LARGE SALAD FOR DINNER, I was alone last night. My wife has been visiting family for a week and, with no one to talk with, it was either TV (which some say sucks our brains out) or reading. Thus, I picked up my copy of *The Adapted Mind*, edited by Barkow, Cosmides, and Tooby (1992), and began reading a highly interesting description of the eye as an example of a complex adapted structure whose functions were fitted to environmental parameters over vast periods of time.

The eye is protected by a transparent cornea that, when scratched, most often repairs itself within a day. The iris, a muscle that surrounds the pupil, responds to light intensity by constricting or dilating the aperture to control brightness. The lens is both curved and transparent and is flexible enough to adjust for distance so that light focuses on the retina. The retina is composed of several layers of neural tissue—**a piece of the brain that migrated to the eye during fetal development.**[2] Groups of rods and cones on the retina translate light, through a series of photochemical reactions, into electrochemical potentials which feed impulses into the optic nerve that guides these impulses to the visual cortex where logic gates sort them for congruence with stored patterns to confirm their nature, identity, and what it means to us.

Logical Informational Patterns

Since nature is all about logical patterns, and mathematics is the language of nature, I wonder if mathematics could be described as the language of logical patterns? A pattern is the logical ordering of particulars in a larger web of relationships that exists as some form of unified whole. The human eye exhibits a highly complex logical specificity. On the first level, the cornea, lens, aqueous humor, retina, and optic nerve must each perform a multitude of highly specialized functions that relate logically and "fit" within the broad network of other specialized functions. One function cannot continually interfere with another. Such conflict would generate negative feedback, pushing the system into irrationality. On the next level, the brain must convert electrical potentials into neurotransmitters that set up other potentials in specific groups of neurons that seek congruence in millions of 'files' to identify the object seen, its distance, its relevance to the observer, and any other specifics about it. For instance, the eye sees "dog." What kind of dog is it, do I know it, who owns it, what is it doing, is it a threat, does it belong there, and what response is called for? The brain runs millions of these scans and analyses every day, automatically evaluating, classifying, determining meaning, and ordering responses.

The logic involved in fitting trillions of cells with systems and processes to function harmoniously, while electrical and chemical agents integrate with trillions more of bacteria, viruses, and other critters that inhabit and aid the body, is nearly unfathomable. Who could begin to catalog all these interconnected processes that share information, and map the feedback loops involved without having information working at cross-purposes? Information is the currency and scaffolding of every logical pattern.

Order, pattern, logical connections, unity, balance, symmetry, efficiency, and form—all are fundamental to the way the universe works. Plato suggested as much when he postulated that "God is a geometer." In 1917, D'Arcy Wentworth Thompson published one of the most overlooked books of modern science, *On Growth and Form*. Thompson follows in the footsteps of Pythagoras and Galileo in suggesting that the morphology of all living things can best be explained in mathematical terms. It is these patterned similarities, varied to accommodate size, function, and environmental parameters that bind all of life in a unified continuum. "I know that in the study of material things, number, order, and position are the threefold clue to exact knowledge; that these three, in the mathematician's hands, furnish the 'first outlines for a sketch of the Universe' ... the perfection of mathematical beauty is such ... that whatsoever is most beautiful and regular is also found to be most useful and excellent."[3] The Logic that built the universe is mathematical. Therefore, the language of mathematics, as Galileo pointed out, is the most accessible way of explaining the universe.

Most scientists bristle at the suggestion that the universe "cares" or is "intentional," perhaps because it suggests an unintelligible operator. Yet, that's what Spinoza implied with his concept of "conatus," and Thompson saw in the mathematical ordering of morphology, i.e., that all things strive to remain in their ordered and balanced state of being. Nearly everything resists depatterning. We refer to this continuation in a balanced state as "homeostasis," or "balanced standing." It is no stretch to say that the kidneys care about the balance of electrolytes and waste products in the blood, and that they act with intentionality to maintain balances. If they go to sleep on the job and stop monitoring and adjusting those balances, we soon feel it. The brain at all times seeks to maintain homeostasis with our environment. It stimulates us with things called feelings that generate both comforts and alarms to cause us to react in appropriate ways either to an oncoming car or dinner and fine wine with our loved ones. Feelings and reactions are logical responses to logical patterns that affect our chances of continuance. Intentions are contained in the patterns of algorithms selected by the environment for outcomes favoring continuance. Illogical algorithms that contain false information are erased with their hardware. Malware destroys its carrier. We are carriers.

We live in the age of algorithms. But this age actually started around AD 825 with the Persian mathematician al-Khwarizmi, who later wrote a book on algebra. "Algorithm" is the Latinized name, al-Khwarizmi. An algorithm is an unambiguous set of steps and processes for solving a problem, or an ordered series of steps to gain a desired outcome. All functional algorithms must exist as a logical web of related and non-conflicted inputs or actions patterned to gain answers, form a product, or initiate action. Each part of an algorithm must stand in logical (syntactic) relationship with every other part for outcomes to be systemically meaningful (semantic). A faulty input can yield disastrous results. Consider the crisis at Boeing over sensors and software on the new 737 Max-8 Dreamliner that caused two disastrous crashes recently. The problem can be summarized as an illogical pattern of feedback (non-truth) that destroyed an entire system. Faulty data from a sensor was fed into software that put both planes into unrecoverable dives. Lives were lost and a multinational corporation is now severely damaged.

Information always has an "about-ness." Red is about color. Tall is about height. Age is about lifespan. Disease is about health. Happy is about emotion. Let's play a Jeopardy-style game. I will state a few pieces of information and you decide what it is about. Here we go:

A saffron-colored leaf wafts earthward under a cerulean sky.
A splash ripples on the lake surface where an insect had just landed.
A scream echoes from a darkened street.
A tear came to his eye as he perused an old photo.

All of the above statements are rich with "about-ness." We can imagine several possible scenarios. Each partial scenario is the first sentence from a novel that could move forward in a thousand directions. Yet, they are partial information and unincorporated with the larger web of narrative. They are "about" a larger story which the mind races to complete.

In like manner, we gather vast quantities of information from our sensory environment all day, every day. Often, it is incomplete information with an about-ness that the brain attempts to fit into a larger narrative—this imaginary person we call "Self." Consider: The forecast calls for rain. Tesla drops 20% on disappointing earnings. Rumors are floating about future layoffs at the plant. Real estate values are rising fast in your town. The price of solar panels dropped by 20% this year.

In each case, the relative value of the information is determined by its impact on us. Each of the above sentences might affect particular persons in very meaningful ways. If the rain, the job, the stock, or the price of solar panels is a relevant factor in your life, then the "about-ness" becomes more critical because it goes to the center of your narrative—yourself and future prospects.

Life and the world exist in webs of relationships where changes in one relationship produce informational feedback that affects other relationships. Thus, relationships, if they are at all meaningful, are characterized by reciprocity. It becomes clearer each day that human activity has an effect on weather, and weather has an effect on us. When connections are mathematically logical the system is intelligible. DNA determines RNA which determines proteins which determine structures and chemical reactions which determine function, health, and continuance. Information is fed back into the loop that regulates structure and function. If the information is faulty, the pattern either rearranges, corrects itself, or disintegrates. Logic is determinative.

Logical informational integrity is basic for continuance. Illogic is destructive absurdity.

Information is the currency of logic. Currency is agency. I will go to the store tomorrow for milk, eggs, bananas, and bread. Yet, my need is an inadequate agent for procurement. I need an agent, a mediator between myself and the people who run the store. I need money, or the sub-agent of money—a valid credit card—because money is stored energy which is a derivative of work. Thus, I could bargain with the store owners to exchange the energy of my work for their groceries, or I could give them the equivalency of that work that I've stored in money. The basis of energy is information sent from the sun via photons that translated to energy on Earth.

Therefore, if information is connected logically, viable patterns emerge naturally. If all of the information about the eye is translated faultlessly from the embryo's DNA, the fetus will develop two perfectly formed and functioning eyes. This multitude of logical patterns, derived from information, makes the world an intelligible place.

Every intelligible thing that exists and continues is a Logical Informational Pattern.

Autopoiesis

In 1972, two Chilean biologists at the University of Santiago, Humberto Maturana and his student Francisco Varela, proposed the *autopoiesis theory* as the process for the emergence of all systems, including life and cognition. Autopoiesis is a compound of two Greek words—*auto* = self, and *poiesis* = making. Autopoietic systems, therefore, are self-creating systems capable of maintenance based on a logical informational pattern that interacts with its environment for both continuance and self-replication.[4]

Emergence is a dividend of autopoiesis. Life and consciousness are emergent properties that cannot be found in any system's constituent parts but emerge from a whole which is greater than the sum of its parts. Disassemble the body of a squirrel, for instance, and spread the parts across a dissection table. You would see piles of muscle, bone, fat, fur, organs, brain and nerves, etc. We could then reduce these to cells of similar type, which could then be broken down into molecules. Not one of these pieces or groups of pieces would suggest the arboreal acrobatics that were a former quality of this assemblage. Nor could we deduce an inkling of the instinct that drove this creature to cut and store acorns every autumn as provender for a coming winter. Life and the qualities of life are emergent properties.

At base, life is a logically patterned informational system constructed of an interconnected web of autopoietic subsystems…of tissues, of cells, of molecules.

Life is a natural product. The English word, nature, is taken from the Latin, *natura*, and means birth, origin. The comparable Greek word is *genesis*. Nature is genesis. Nature is birth, fecundity, the emergence of higher qualities from logical information in diverse and novel patterns shaped by interactions with the environment. Our uniqueness as persons is due to the diverse patterns of information and matter that make up each self and each body. We are varieties, each somewhat different, resulting from information's fit with existing realities. But what drives autopoiesis?

Basins of Attraction

The Town of Watkins Glen lies at the southern tip of Seneca Lake in Central NY State. The "Watkins" part is in honor of a Dr. Watkins who was beloved by the town. "Glen" is the result of an ancient seabed from the Devonian period that was uplifted, fractured, and eroded by meltwater from glaciers of the last Ice Age. It is worth spending an afternoon ascending 500 feet of elevation on the stone walkway that parallels the existing stream and waterfalls. One walks this chasm in deep time (375,000,000 years' worth) amid fossils, stratified stone layers left from the bottom of ancient seas, mosses and ferns clinging to moist walls of stone, water thundering its way to the sea, while immersed in a moist, chthonic breath rising from the bowels of Earth.

The observant student of Earth will notice, as she hikes ever upward along the stream, a series of deep potholes carved randomly in the stream bed by the unceasing scouring of flowing waters. Water jostles its way downward and temporarily comes to rest in these deep bowls of carved stone, only to be ejected by more water racing downward. These are what might be called *Basins of Attraction* because they are, temporarily, the lowest point to which gravity can attract the water. For instance, take one of those aluminum saucers children use for sledding on snow. Now throw a ball bearing along the inner rim such that it circulates inside the edge. Gradually, as the steel ball loses energy, it will cycle ever downward and eventually come to rest at the center of the saucer. That center is a basin of attraction because it is the lowest energy level and most efficient state the steel ball can achieve and remain stable. Or think of a pinball machine. The steel ball enters at a higher energy level, but ultimately ends up in the basin behind the flippers because that is the lowest energy point and ultimate basin of attraction.

Why is this important? Because it explains why life and the universe are the way they are. Our solar system formed the way it did because of gravitational basins of attraction where matter could settle at its lowest sustainable energy level and exist in balance with all the other basins of attraction around it. When forces get jiggled or out of balance, new basins of attraction will form. The jostling waters racing down Watkins Glen continually knock water from higher to lower potholes, then to Seneca Lake, to Lake Ontario, to the St. Lawrence River, and to the Atlantic Ocean as the ultimate basin at the lowest energy level. Thus, an entire system creates itself simply by seeking gravitational basins of attraction. Scientists are now in reasonable agreement that this is the explanation behind life and evolution—systems create themselves (autopoiesis = self-making) based on attractions to the lowest sustainable energy levels, which are also the most efficient position for long-term continuance.

Food digests because enzyme/substrate complexes are an electrostatic basin. Lake Winnipesaukee is filled with water because the 104.5° hydrogen bond angle with oxygen is the lowest sustainable electromagnetic basin for water molecules, and the lake basin is the lowest temporary gravitational attractor. Oxygen in the lungs is drawn by electro-chemical attraction to hemoglobin, which then loses that oxygen to myoglobin, which is a stronger attractor, in muscle tissue. But then, hemoglobin becomes an attractor for CO_2 until it is attracted away in the lungs. Thus, our every breath rides on basins of attraction. Long-term memory is formed by the connection of encoded bits joined at their lowest, most efficient energy levels and can be called up by associations with any one of those other basins which are all linked together in our great neural pothole, just as water in the St. Lawrence River is linked to the Atlantic Ocean and Watkins Glen. As water links all parts of a drainage system, connections of neural and extra-neural basins unite all of our stored memories in logically connected patterns.

In the macroeconomic view, this explains why higher energy requirements such as private health insurance and fossil fuels cannot survive. By their rent-seeking financial (energy) burdens, they have become forced inefficiencies that do damage to the system and interrupt the inexorable pull to the lowest energy, highest efficiency basin of attraction of less expensive renewable energy and a citizen's health care coop. When politically protected inefficiencies that extract energy (money) without adding proportional value are removed, free-market capitalism will always seek the lowest energy, highest efficiency basin of attraction. At that point, the system achieves optimal function for long-term sustainability and ultimate benefit to all members of the system (longer, healthier lives, cleaner environment, reduced threat of global warming). Electric utilities bought more sustainably generated electricity than coal-fired electricity in April of 2020 because it was cheaper. People in other developed nations with single-payer health care live longer than Americans at approximately half the health-care costs.

Dan's Incompleteness Hypothesis

All four primary forces demonstrate *attraction*. By their very nature these fields affect a clumping of matter, energy, and information by attractive force/field. Silicon dioxide molecules are attracted to each other and bind into logical crystalline structures based on electrostatic symmetry. Hydrogen and oxygen naturally bind into water in three phases. Bacteria clump into thin films for survival and reproduction. Starlings and fish aggregate into flocks and schools that move as one. Due to its field nature, information will accelerate technology toward super-algorithms, quantum computing, and machine learning. Matter compresses into stars, and stars into galaxies and galactic systems. Ants and bees aggregate into colonies. People unite in families, tribes, and nations. Logically incomplete pieces of information are agglomerated by the agency of attractive forces into completed patterns.

Attraction to greater concentrations of the same type provides reciprocal benefits. A single ant or bee by itself is highly vulnerable and has little chance of long-term survival. Fish and fowl clumped together more easily confuse predators—a calculation to reduce risk.[5] Man and woman join to increase and extend themselves into the future. Social, vocational, and ideological groups collectivize for a host of reasons obvious to any reader. Universities attract students because the dense informational field provides a broader menu for learning which enhances life opportunity and power. Co-ops can negotiate better pricing. Grouping, concentrating, adhering to alliances, and fusing into larger units provides emergent qualities and benefits of scale unavailable to individual units. One taxpayer cannot build an interstate highway system.

In this regard, it seems to me that Spinoza is incomplete without Nietzsche. Spinoza defined the basic drive of all things as *conatus*, the striving to persevere in present form. Nietzsche famously taught his *Will to Power* of the Uberman. Observation indicates that understanding of the universe comes by combining these doctrines. The four forces do indeed hold things together for continuance. But they also aggregate mass and power relentlessly. Consider the gravitational field. Our solar system was formed from the remnants of stellar debris concentrated by gravity to form sun, planets, and moons. And the process continues. Any matter that gets too close to our gravitational field will be drawn in to merge and increase the substance of earth, like the asteroid of 66 million years ago that decimated 90% of life. Consider this question: Is gravity ever satisfied? Theory teaches that most systems seek homeostasis, or a balanced state. But we seldom reach that state because of the force of attraction. The bigger the gravitational mass, the more power it has to pull other matter to itself. Matter, energy, and information continually combine into novel and larger units because of attractive force. Therefore, I conclude that the four forces drive the universe to continue and to concentrate power. Force fields are greedy. Black holes suck up light.

Other factors, however, indicate that growth cannot be limitless, at least here on earth. The redwood groves of northern California are nature's cathedrals. Their massive trunks rise like fluted columns supporting the vault of heaven. Silence. Peace. Harmony between deer ferns, banana slugs, and the condensed mists rolling off the Pacific. They seem immortal. Disease, fires, browsing by animals—none of these seems to affect them. But gravity does. The number one natural cause of redwood mortality is falling over.

Likewise, animal populations can only go so far in Earth's closed system due to thermodynamic limits. Without limits, the wetlands in front of my cabin would be full of beaver. But that won't happen because there is only so much new vegetation for them to eat each year. At a certain point, overpopulation demands more calories than the system can sustainably produce. Therefore, a few beavers become calories for predators, two-year-olds are forced out, and when the food supply dwindles, the entire colony moves on to greener ponds. This same process must eventually play out in human populations which are currently out of balance with the natural ecosystemic services of Earth.

Even in limitations, Logic prevails. Informational feedbacks flash warning signs of imbalance which is illogic, and the pattern bifurcates into new orders that achieve homeostasis at a lower organizational level. Pieces and parts of former systems are attracted to re-pattern at the lowest logical basin of attraction. All things are subject to this adaptive cycle. Repatterning is a generative force issuing in novel forms that can sustainably integrate with present environmental affordances. Chaos is creative. Destruction fertilizes genesis.

Reciprocity between the greater Earth and its creatures is the basis of belonging. Everything that happens has a reciprocal benefit, from photosynthesis and respiration to the hydrologic cycle to the geomorphic features of Earth. Our mortal bodies rise out of the Earth and return to it. We are the frogs and trees and clouds, the stuff of stardust flowing from the cornucopia of creation. It is this attractive reciprocity that fuses multiple bits of information, molecules of iron, carbon, calcium, oxygen, hydrogen, hearts and lungs, man and woman, people and animals, groups and tribes into an emergence that surpasses anything we might have imagined. And it is all "autopoietic" based on the logical agency of natural forces settling in the most efficient basins of attraction.

But why this continual striving to continue and increase? If the

Logic (God) running the universe is complete, lacking nothing—omnipotent, omniscient, and omnipresent—why are its agents (four forces) forever greedy? Why must novel forms continually arise while others enlarge by attraction, disintegrate, and emerge again as variant forms? If Logic is perfect, why can't it stop building? Does this imply that God (Meta-Logic) is incomplete?

Yes and no. Yes, from our perspective because we can never know all the probabilities in the quantum wave function, only those few selected by observation. But if we had unlimited time, we could actualize every probability. Meta-Logic, however, is not governed by time. It is timeless or stands outside of time. Therefore, since we are time-dependent, we see only probabilities as revealed at points in time. On the other hand, in timelessness all things appear as present reality. Thus, God is complete, but we do not see that completeness because we are at one point on time's continuum.

Language is Logically Patterned Information

The general understanding of language (do you speak French, English, Mandarin?) is severely limited and falls short. Come, walk these labyrinthine forest paths with me and bathe in a language that is multi-sensory—a language that communicates information clearly through the nose, eye, ear, feet and hands, tongue, and spirit. The language of the universe is 13.8 billion years older than human words and has been coursing back and forth between stars and galaxies, rocks and redwoods long before humans took a stick and began making marks on clay tablets.

Three pairs of geese are back in the wet meadow. They were

standing on the edge of the yet ice-covered pond yesterday and one could almost hear them tapping their feet impatiently, waiting for the ice to melt and provide them access to fresh food. One male closely followed his smaller "wife," nudging, touching her neck, hovering as if to say, "Honey, it's time and I need a little relief." All of this is clear language because it communicates understanding. Language is intelligible patterned information passed from one medium to another to yield meaning.

Information oozes from Earth and Earth systems at every turn. The wood ducks' repeated flights past the nesting box I attached to a tree at the pond edge tells me something. Fresh turkey tracks in the sand along the road all point to a neighbor's house and speak of hunger and food and a bird feeder. Swollen buds, daffodil leaves poking through the ice, snow sliding from the cabin roof, a fallen tree, the smell of ferns…everything conveys information. When that information is logically placed in pattern with other information it is syntactically intelligible. The above information conveys volumes of meaning about the season and condition of life in this little kingdom; it tells who is doing what and why, and how that might impact my life and prospects.

Human language is simply a matter of encoding information about one system into intermediary and abstract symbols (metaphors) which are reinterpreted as electrical potentials that convey information to the brain and body to achieve communication (common union = understanding). The brain likes picture-patterns. Metaphors are picture-patterns. Language is a logical metaphor with symbols in their proper place.

Philosophy's Vision

Philosophy—the love of wisdom—has long sought to understand the reality behind appearances. Plato did this with his metaphor of a cave where uncritical viewers conflated reality with shadows moving on a wall. Then he castigated his audience, and humanity in general, because we judge too often by appearances. No, says the sage, reality is found in the unseen actors moving in front of flickering torches that cast the shadows on the wall that you regard as "the real." Likewise, he goes on, everything on Earth is a shadow of real, or ideal, forms that exist in heaven. Every oak tree on Earth is a flickering shadow of the ideal oak existing beyond the realm of Earth and vision. Although this obvious dualism has gone the way of the Loch Ness Monster among the philosophic community, it may yet hold a degree of truth.

The world entered a new period in 1948 with the publication of Claude Shannon's, *A Mathematical Theory of Communication*. Using Boolean algebra to digitize information and building on Norbert Wiener's emerging discipline of cybernetics, particularly *Probability Theory*, Shannon opened the door to *The Information Age*. With our near total dependence on digital devices hooked to the internet and powered by algorithmic apps and search engines, we are all like leaves dangling on the tree that sprouted from Shannon's seed. We spend our days in awe of pixels flickering on a screen much as Plato's friends were entranced by the image of shadows in a cave. We continue to conflate reality.

Shortly after Shannon opened the gates, the life sciences began moving toward an informational-theoretic foundation with the analysis of DNA's double helix structure. This became the basis for understanding a multitude of interrelated forms manifest as life. From here, innovation evolved to genetic algorithms, bio-computers, ecosystemic energy flows, gene editing, the computational basis for quorum sensing, cognitive sciences, and neural networks. In the process it seems almost as though we've returned to Plato's conception of reality.

Below the symmetry and order of the universe, at the level of daily life, the world is often a messy, fickle, random, and chaotic place. Multitudes of beings strive for conatus, often at cross-purposes with one another. Small perturbations in meteorological systems wreak havoc by hurricanes, floods, and tornadoes. The assassination of an archduke precipitated a global conflagration. Genes are damaged by chemicals and radiation. A mother drinks too much alcohol. Patterns get twisted, marred, and broken. The world is an imperfect place.

So, perhaps Plato was onto something with his ideal forms. As was Aristotle with his divine cosmic principle that makes the universe intelligible. For my money, all those ideal forms and that cosmic principle are summed up in the concept of *Logical Informational Patterns*. These are the ideal forms, the logical principles that make the world intelligible. Informational patterns that have adapted to and been selected by the affordances of their environment over long periods of time are the fruit of pure logic, selected by thermodynamics. Therefore, I agree with Plato that there are ideal forms not manifest to the eye. But the ideal form is in the evolutionary code honed over eons of time to fit hand in glove with environmental parameters. There are no ideal forms locked away somewhere in the vault of heaven. They exist, unseen and immaterial, within the logic of their patterned information right here on Earth. We can observe and measure the carriers of that logic and information such as DNA. And we are near to understanding the logic of how genes structure proteins and proteins interact with substrates; of how methylation suppresses gene expression and how cancer fools the immune system—a whole new language of genetics, chemical semiotics, feedback, and epigenetics. We are at the point of putting logic under the microscope. This is huge, because logic is causal via the agency of forces manipulating information. Cancer is a logic problem. We fix the form by fixing the logic, whether it is a diseased body, a failed religion, false economic system, or a sick culture. Logic is fundamental. Logically integrated information is health and life.

The Earth today seems no more interested in my wishes than it is in the date on the calendar which indicates that we are well along into spring. Three inches of snow settled softly through the hemlocks yesterday. The roads are sand and slop. Visibility across the marsh is now about a half mile with the mountains occluded in a mist of clouds. The geese continue in their frustration, pacing about the iced-in pond waiting, waiting... The logical pattern of the seasons seems to be breaking down.

It would be easy for me to conclude that, *aha*, global climate change is playing out on this present stage. But I don't know that. Perhaps the relentlessness of this winter is just another perturbation of the system that will even itself out in another week. But maybe it won't. Yet, it seems that as our data bank of climatic information grows, and as atmospheric carbon dioxide concentrations rise, perturbations of several types are increasing in frequency. The *Uncertainty Principle* appears to be valid. In the end, we're all gamblers betting every day on probabilities—the stock market, driving, flying, taking a shower, eating bacon. The waterfowl out in the marsh are betting on probabilities; the beavers are betting on probabilities; the deer are betting on probabilities. They are betting that spring will come, green grass will grow, offspring will open their eyes to a new world in May. Life is a wager that logic will prevail.

4. Mind

"The criteria of mind that seem to me to work together to supply
this solution are...the phenomena which we call thought, evolution,
ecology, life, learning, and the like..."
— Gregory Bateson[1]

"There is an aspect of the modern 'cultural' mind that...shows an
alarming degree of dissociation and psychological confusion. We
believe exclusively in consciousness and free will, and are no longer
aware of the powers that control us..."
— CG Jung[2]

Walk With Me

I LIKE CITIES WELL ENOUGH, in small doses hyphenated by long
periods of time. Cities are living things, and like all living things they
get on by processes. Information from those processes is everywhere
to stimulate one's mind via the senses. Horns honking, eyes screening
a multitude of faces for feeling, insight, and reflection, the smell of
diesel smoke, bagels cooking, the hardness of the pavement pushing
against our feet, the feel of a park bench, pigeons scavenging, window
shopping, the museums, and theaters—all make a trip to the city a
multi-sensory feast of information. Cities are systems nested within
systems, thus minds nested within minds. To enter any system is to

enter another mind full of logic and patterns and information that interact with our own systems of logic and patterns and information so that there is a cross-fertilization of both. Each stimulates, rearranges, and enriches the other.

A walk with nature is a union of minds, one mind sensing and remembering (albeit subconsciously) its own reflected development within a greater mind. The mind of this forest and wet meadow is a repository for mental prototypes of what I call "my" mind. But if "my" mind flowed from all of this, how can I say it is mine? Is it not ultimately the mind of Earth, patterned by the Logic of the universe, that exists in this bipedal carrier who thinks he is the pinnacle of creation? How could it be any different? Mind enfolds "me" in a multitude of developmental stages like the deep breath of atmosphere. Everything about the kingdom of nature speaks to mindedness.

Weather is an endless conversation between Earth and sun, defined by a syntax of heat and cold, wind and rain, pressure systems, and angle of exposure that issue in the semantics of the seasons. Information riding on photons transmutes to multi-variant forms in melting ice, thermal currents caressing the side of an ancient volcano, the informing of energy within chloroplasts of leaves that build acorns that built the body of our bear now rousing herself to find new cubs suckling those transformed photons from her breasts.

The sun and trees carry on a different conversation with each season. Six weeks ago, the sun passed information to the maples that it was time to prepare the nursery for nascent leaves. The sugar maples responded by opening valves for sap to flow upward from root to crown, to branchlets and buds that began to fatten. It was also a signal to New Hampshire countrymen to drill and tap those trees for New

England ambrosia—maple syrup. But the sugar-bushmen are not the only ones tapping trees.

Walking the path from cabin to house yesterday I found a massive pile of shredded wood and bark at the base of a tree. Pileated woodpeckers are back, and they are hungry, especially since the female needs calories to produce healthy eggs from which the next generation will soon emerge. A dead white birch has stood along the path for three years—long enough to attract a multitude of insects that laid eggs that became grubs that feed woodpeckers. It's the perfect example of the circular logic of the forest. Energy and information flow from sun to tree to insects to woodpeckers to Earth to tree to insects to woodpeckers in an endless cycle. At each stage difference that makes a difference is selected by mind as the causal basis for decisive action. The entire system is mind. What we see are only vehicles of mental process. Logic is everything expressing itself in selectively patterned systems we call minds.

To walk in the woods, the city, a factory, a school, a business, the senate, the grocery store...is to walk in mind. The logic of mind is the reality that generates all the illusory images flickering across the walls of our various caves.

An unseen cast of encoded actors calls the shots to create the flickering image I call a forest. They are nucleic acids, proteins, hormones, auxins, photoreceptors, symbiotic relationships between bacteria, fungi, root tips, soil, neurotransmitters—adapted patterns coded into minds and bodies across eons of evolutionary time to survive and reproduce. Yet, these are also only sub-agents causing the bear to hibernate and the pileated woodpecker to shred the birch in her search for grubs. Other agents stand behind these—the four forces of the universe that developed a multitude of novel arrangements for thermodynamics to select and drive evolution. And behind these, pulling the strings like deus ex machina, stands Logic Deus. **Meta-Logic is Ultimate Causality expressed as Cosmic Mind.**

Cosmic Mind

By its very nature, philosophy is driven to find the fountainhead, the source, the cause behind everything. Perhaps that's because humans want to "make sense" of life and make the universe intelligible. For, if the universe is not intelligible then we don't know what the world is about or why we are here or what purpose, if any, there is to life. Without intelligibility we have no anchorage. Life and the world, and ourselves in particular, would have no more meaning than a piece of flotsam drifting endlessly on a meaningless ocean because it all came about by random impersonal processes that have no particular point in the icy cold blackness of space, where nothing is related to anything. That's as good a definition of hell as I can muster, and it is depressing. And depression is a very bad place to be. People kill themselves or waste away in depression. Life and progress require a rational basis. That's what philosophy seeks. Irrationality is the insanity that destroys systems.

Order is the antidote for chaos. Reason is the response for confusion. Truth vanquishes falsehood. Knowledge fills the empty cup of ignorance. These are the things that have motivated philosophers in every period. They motivate mathematicians and physicists today, just as they motivated Bohr and Planck, Spinoza, Descartes, Kepler, Copernicus, and the ancient Greek philosophers who mocked St. Paul.

They were all attempting to make the universe and life intelligible, to provide a basis for understanding—an examined life—a trustworthy basis for knowing something.

Anaxagoras (510-428 BC) was both a Persian and a Greek citizen. He served in the Persian army and later, to pursue his philosophic and scientific interests, moved to Iona. He appears to have been the first scientist to give an accurate explanation for eclipses. Anaxagoras' highest contribution to philosophy was his concept of *Cosmic Mind*. He believed that mind (universal logic) was the ordering principle of the universe—mind linking perception and creation. Socrates was strongly influenced by this concept.

In Plato's philosophy, the soul is divided into three-parts. Reason is given the most excellent part, an independent realm of pure thought whose object is always the "idea." Reason, because it functions in the medium of knowledge, controls moral decisions just as the helmsman controls a ship. The zenith of Plato's insight is that God's consciousness abides in reason. Thus, reason is the highest path to authentic humanity. Perhaps he was echoing Euripides' earlier writing, "For the mind in each of us is the mind of God" (my translation).

Likewise, the old Stoic, Zeno, described God as cosmic reason, in another echo of Anaxagoras. Similarly, Epictetus said God's being is "Mind." These concepts filtered down to a Hellenistic Jew, Philo of Alexandria. Philo lived at the time of Christ and devoted considerable effort to harmonizing the Talmud with Greek philosophy. Scholars believe he strongly influenced the Johannine concept of "Logos" as the creative principle of the universe. Philo wrote that each heavenly being, each star, is pure Mind. God Himself is pure mind in the fullest and final analysis.[3]

If the basis of mind is logic, and logic computes informational potentials relative to the system it governs, then logic needs a way to evaluate those potentials, just as mental patterns evaluate information to determine voltage potentials for neural firing. In order to evaluate, therefore, one needs standards of valuation. Value is measured by the evolutionary axiom that asks, "How favorable is this thing, event, possibility, news, etc., to self-perpetuation and transmission of coded information into the future?" Evolution endowed the brain with the capacity to select and store knowable environmental patterns that impact organisms and assigned a value, or valence, relative to survival potential. Feeling, behavior, and culture are effects of valence relative to survival and reproduction. Stored neural patterns, therefore, are standards of evaluation to determine life probabilities and the best decisions. Since no one ever has all the information about anything, the best choice is computed for probability potential. For example, you are watching traffic from the sidewalk, waiting to cross a busy street. You observe a break in traffic and decide to cross the street. That decision was a massive instantaneous computation involving distance across the street, distance to the closest oncoming car, speed of the car, speed of walking/running, and the probability that you judged all factors correctly. But were you certain of all those factors? Much of life is risk management.

And the goal of that management is the continuation and expansion of Cosmic Mind in temporal patterns that reflect varying facets of that unified Mind.

Participatory Universe

We've come at the topic of mind from the perspective of ancient philosophy, but we live in the age of quantum mechanics. Does contemporary physics shed any light on the nature of mind and reality?

Note: Do not be threatened by the following discussion. Nobody can fully explain the quantum wave function, or how information is selected from it.

In describing the quantum wave function Werner Heisenberg appealed to Aristotle's concept of *potentia,* which he called *objective tendency* for something to occur. Once the wave function is observed, or selected at a certain point, it collapses to reveal the information potential or probability at that point. Since the quantum wave function contains all the information about the system, it carries an optimal description of all available probabilities—Aristotle's potentia.

Henry Stapp is a theoretical physicist at the Lawrence Berkeley Laboratory and author of *Mindful Universe: Quantum Mechanics and the Participating Observer* (2007). Reiterating some of his main ideas in a subsequent article, Stapp says,

> This conception of the quantum brain is intuitively accessible, and it is made possible by (environment-induced) decoherence ... (the) brain is an evolving cloud of essentially classically conceivable potentialities ... Each human experience is an aspect of a psycho-physical event whose ... physically described aspect is the reduction of the cloud of potentialities to those that contain the neural correlate of that experience.[4]

If I am reading correctly, Stapp is saying that our lived experience is like a quantum cloud. Information or situations of interest to the observer allow that person to question the value of that information relative to survival and reproductive goals. We do this by selecting information contained in the wave function and comparing that with stored neural patterns which serve as standards for comparison to compute + or - valence and degree. We can select information at certain points in the quantum wave function that are relative to our interests and compute them with neural patterns for subjective value.

This computation permits us to make decisions based on probable outcomes. Choice, both conscious and unconscious, is reciprocally determined by our subjective reaction to objective factors in the environment.

The clear implication, however, is that the universe is participatory, that we are not predetermined robots dragging across the stage of time. We make choices based on probable outcomes, like crossing that street. The universe is an often wild, chaotic, and indeterminate place to which the individual must respond and adapt based on the fundamental axioms of evolution. In the process, life, mind, and logic build themselves forward in ever increasing complexity and beauty. Ultimately, the environment designs and builds us, because our responses must fit with its reasons. *Deus, sive Natura.* Nature is in the design/build business.

The Helmsman

A new science of cybernetics began to develop during the 1940s under the leadership of Norbert Wiener (1894-1964). A precocious child, Wiener was homeschooled by his father and went on to earn his PhD from Harvard at age 17. He spent most of his life teaching mathematics and philosophy at MIT. During World War II he worked for the government on guidance systems for anti-aircraft guns and missiles, using his keen mathematical aptitude to plot the interacting complexities of arc, velocity, and trajectory to improve the accuracy of our weapons systems. During this period, he developed his concept of the "feedback mechanism" as vital to the guidance of projectiles. Since systems are dynamic, conditions are always changing. In order to be successful, the system must continually adapt to all those external variants. Each part of a system is impacted by change and feeds information about that change into the larger system as feedback. Based on feedback from components, the overall system computes values relative to success. Nature has been doing this for 13.8 billion years.

Wiener put his ideas into a 1948 book entitled, *Cybernetics*. This is the word Plato used to describe government, and comes from the Greek word *kubernesis*, which is most often translated "helmsman." It was a common word in the Mediterranean world of the Greeks, who had been sailing Homer's sea for a thousand years. They understood how critical the role of the helmsman was in guiding the ship to safe passage. The most dangerous point of Odysseus' return to Ithaca was the night passage between Scylla and Charybdis, one a multi-headed monster that devoured sailors and the other a powerful whirlpool that sucked ships to Neptune's chamber. Everything depended on the helmsman getting true information and making logical decisions based on that information. The helmsman must steer the middle way—the knife-edged ridge path. On each side lurked disaster.

Gregory Bateson

Some time ago, I found myself living in The Villages, Florida—a place of mass-built houses, concrete, roads, shopping plazas by the dozen, postage stamp lawns, with nearly every inch of Earth sprayed with pesticides. Feeling caged, I wondered if I was losing my mind. But this suffering had an upside: I began a search to understand what is mind, where is mind, how does mind work? It became clearer that the boy who had grown up in the forest and mountains had integrated those elements as part of his mental frame. To be cut off from nature, therefore, was alienation—separation from the womb of mind, hence, "out of mind." This search led me to Gregory Bateson (1904-1980) and his *Mind and Nature: A Necessary Unity*, and then to a larger synthesis by Noel G. Charleton, *Understanding Gregory Bateson: Mind, Beauty, and the Sacred Earth*. The latter book is more accessible. Gregory Bateson is one of the top five influences for my worldview.

Gregory's father, William Bateson, was the famous geneticist who named the new discipline in 1905. He named his son in honor of Gregor Mendel. Gregory's grandfather, W.H. Bateson, had been Master of St. John's, Cambridge. It was there that Gregory took a degree in biology. Over the course of years, Bateson became a polymath and later served as a Regent for the University of California. In 1935 he married Margaret Meade and together they did anthropological research in the South Pacific. After the war Bateson turned his interests to the new field of cybernetics, with Norbert Wiener, Walter Pitts, John von Neumann, et al. For seven years they worked on the theme, *Feedback Mechanisms and Circular Causal Systems in Biological and Social Systems*. From his research, Bateson concluded that the **system is the mind.**

Mind exists, Bateson said, wherever action is taken based on recognition of a difference that makes a difference. Kidneys have mental properties. Bacteria in our gut comprise a mental network capable of making decisions based on information (difference). Political systems, social groups, religion, biological communities, your cat, families, the lake...are all mental systems—perhaps not consciously, but minds nonetheless. Communication and control are decentralized in mind (the brain is not the mind), which is distributed through recursive sub-systems, each offering feedback that is weighted for its ability to affect total systemic balance. False information, toxins, or new imbalances introduce chaos into the system and may cause it to fork in a new direction and return to balance at another equilibrium point. Imbalances manifest as illness—mental, physical, social. In an aquatic system frogs and fish die as algae increase. Political imbalance is manifest in polarization, distrust, protests, riots, coups, and wars. Businesses adapt to new market realities or die. Changing heat levels in the atmosphere lead to glacial melt or desertification, with accompanying population crashes. The point is, the mind is unbiased and unemotional—its only goal is equilibrium within environmental parameters. Bateson (an atheist) counseled against messing with the "Ecological God," thus elevating the integrated mind of Earth to the role of deity.

From his reflection on mind, Bateson came to the conclusion that conscious human purpose, "might contain systematic distortions of view which, when implemented by modern technology, become destructive of the balances between man, human society, and the ecosystem of the planet." He saw dualism as a terrible error, repeatedly arguing that everything is interconnected. Because our purposes are linear, from need to goal, we either disregard or are ignorant of multiple systemic consequences of our choices that may cause negative feedback and destabilization. Heideggar, Weber, Malthus, Ellul, Ehrlich, and Bostrom all came to the same conclusion. Near the end of his life Bateson concluded that only a religious-type response with a new epistemology could save us.

Brain and Mind

No structure in the known universe comes remotely close to the complexity of the human brain. The connecting fibers in our brain would encircle the Earth 40 times (1,000,000 miles). Even a mouse brain has 450 meters of dendrites and one to two kilometers of axons/mm^3. By comparison, our most powerful computers are like a child's toy. How was it possible for time and endless experimentation to network 100 billion neurons and trillions of connections capable of a quadrillion calculations/second within a three-pound structure composed of 60% fat, plus a multitude of proteins, connective tissue, trace metals, and neurotransmitters that can build an internet, send a man to the moon, cast thought beyond the edge of the universe, and conceive of black holes? Is this what Logic does to reflect itself, or is it a crap-shoot outcome?

The most frequent responses to that question are either an appeal to a personal, all-knowing, all powerful anthropic God who needed someone to talk with, or we are a chance accident of matter interacting randomly across 13.8 billion years. I propose a third way—immaterial Logic working through the agency of the four forces of the universe to pattern matter, energy, and information into logically structured forms selected for integrity of fit (adaptations) with environmental affordances to enable their continuance. Some may ask where the Logic and the four forces came from? My answer is that of Aristotle and Aquinas—they exist of necessity. The nature of nature is because it is. Every religious and philosophic system is built on the premise of necessitarianism. Why is there something rather than nothing? Out of necessity.

Why did Logic and the forces of the universe build the human brain? I say, because it is a logical necessity for Logic to manifest itself in every *potentia* in order to fulfill itself. Unfulfilled potential infers imperfection, or lack, just as human limitation implies that there is a point where we run out of words to express a thought—ineffable. Since it is perfect, Meta-Logic must continue to fulfill every form of novel combinations to manifest itself until it is all in all—a Singularity (Pure Unified Consciousness) where potential is impossible. This is also a logical necessity. It is the nature of Nature to strive for perfect union. The clear implication is that imperfect humans are not the pinnacle of evolution.

But the brain is not mind. Mind is what the brain does. Brain, that yellow glob of fat within our crania, is like the hardware of a calculator. Mind is a logically patterned and immaterial informational system (software programs) capable of discerning differences that make a difference, that can make a decision based on the survival value of brain's probability calculations relative to that difference and effect some action designed for optimal life continuance. And mind did not begin with homo sapiens.

Viruses known as bacteriophages (eaters of bacteria) attach to a bacteria, dissolve a hole in its membrane, and inject their genetic material for the purpose of commandeering bacterial replication, thus turning a bacterium into a virus factory. Over time, however, bacteria have learned a few tricks. Many have learned to recognize viral genes upon entry. They immediately release a protein to snip sections of viral genes and insert genetic malware that destroys viral nucleic acids. This is a learned response that is then passed on to other bacteria. Thus, we can only conclude that bacteria are mental systems because they can learn, remember, and make probability choices based on informational differences.

Steven Pinker believes mind is information processing (computation). And that "information and computation reside in patterns of data and in relations of logic that are independent of the physical medium that carries them."[5] He's simply agreeing with Alan Turing's computational theory of mind which says that information is logically patterned code that exists in patterns of electrical potentials housed in neural networks.

All of this removes mind from the sole domain of man. Previously, we thought that the world of nature—all those unthinking, unfeeling trees, grass, snakes, birds, frogs, crusty lichens clinging to a rock, and stars circling aimlessly above—was just neutral stuff, decorative treatments, resources placed here purely for our enjoyment, use, and extraction. The world of nature was cognitively neutral—just matter.

Yet, as we go beneath the surface, we find a world alive with relationships, webs of mutual interaction, intelligence and cognition, and full of the will to continue. Mind is everywhere—in trees and birds that know the time by the angle of the sun, waterfowl that know where and how to navigate and return and bond with a lifetime mate, my bear that knows when and how to drastically slow her metabolism, snowshoe hares that are super-aware of the bobcat lurking in the balsam thicket, spring peepers that know what to do by the water temperature... The list is truly infinite. And all of this is mind doing what mind does—computing, choosing, and adapting.

All the pieces and parts that formed my cognitive system exist right out there in the wetlands and pond. The only differences are in volume and network sophistication. Our neurons and neurotransmitters work the same way. The trees have signaling and computing capacity functioning through chemical messengers that exist with a network of bacteria and fungi that pass chemical and electrical responses back and forth within the greater forest all about. Birds and squirrels, trees and plants exist in a mutually beneficial network of relationships to share materials and information for the continuation of being.

Therefore, perhaps when we walk in forest and field and sense synergism—a consanguinity—it is because we are all related. **Is not every living mental system that exists in nature a simpler prototypical iteration of ourselves?** We identify with nature because we are nature. Evolution is a mental process.

Mindedness is a property of this broad Earth and its inhabitants. We merely partake in it, one piece of a greater system. How could we ever imagine that mind, Earth, and God are segregated into some dualistic hierarchy? Earth, life, and the stars are one. The universe is a Monism. "*Deus, sive Natura,*" says the smiling Spinoza.

All living things, therefore, are informational systems through which matter flows. We are persons not on the basis of our temporal bodies, but in the information and intelligence we carry. Our personhood is our story. There are 7.4 billion human bodies walking about the Earth today. We are all, more or less, the same model with slightly different trim and apps. It is the story that makes each one unique. We are our story—our stored sensory perceptions, experiences, and memetic software. The true "You" is an embedded story.

Ratcheting to another level, all business, culture, religions, institutions, government, universities, and ideologies are informational systems through which matter, services, and/or information flow. These minds, like those of nature, must evolve, change, and respond to continually changing environmental affordances, reprogram themselves and adapt, or they will be de-selected by nature or the market. When logical informational patterns become fixed and inflexible, they no longer find a reciprocal congruence with ever-changing reality. Incongruence is illogic. They can no longer cope with the world as it is. They become like the lost hiker in the desert, without food or water.

Panarchy

My linear purpose often looks out over our lawn and garden and thinks, "What needs taken care of today?" You know the drill—the grass needs to be cut, the squeaky door oiled, some weeds pulled… But never once have I looked across this vast wet meadow and surrounding forest and asked that question, because it would be absurd. These mountains have stood for 100 million years, beavers have inhabited this meadow, off and on, for millennia, rains run to the sea as ever, and the sun also rises…all without my help. In the natural realm, everything affects everything else with feedback inputs that, with no central authority or decision maker, determine the course of the system based on the laws of nature (Greek, *phusis* = physics). This form of government is called *panarchy* (Greek, *pan* = all; *arche* = to rule or govern). At this point, having lasted 4.5 billion years, it appears to be the best game in town. Why is that?

Command and control (cybernetics) is an emergent property of all complex systems. Complex living systems survive only so long as homeostasis is maintained between all sub-systems of sub-systems composed of tissues and cells and molecules of atoms. Feedbacks communicate temperature changes, water balances, external threats, nutrient changes, toxins, invading viruses and bacteria, emotional states, etc. Based on its own internal needs, the system computes the implication of all feedback and makes adjustments based on statistical probabilities for outcome. There is no "decider-in-chief," just as there is no central switchboard in the brain. Think about it… How many times have you told your kidneys to be careful about the electrolytes in your blood? Or, checked on acetylcholine levels in your brain? Did you make a print-out today for the bacterial balance in your gut? The body is panarchist, just as all of nature is, from ocean circulation to the balance between wolves and elk in Yellowstone.

As humans, we are familiar with many different forms of government—pure democracy, representative democracy, socialist, communist, monarchist, despotic, oligarchic, and even sometimes, anarchist. How do we judge which is best? I rather like Abraham Lincoln's take on the matter when he linked a "new birth of freedom" with a government of the people, by the people, and for the people. Combine this with the Jeffersonian, "life, liberty, and the pursuit of happiness," and the upshot is that organisms cannot flourish apart from the maintenance of optimal balance between each subordinate venue. Therefore, in one way or another, the goal of all forms of human government should be to mimic that of complex systems in nature whose survival is inextricably linked to the well-being of *all* supporting parts, whose disposition is clearly communicated via feedback. When the helmsman disregards honest feedback, the entire ship is endangered.

Now, consider Pope Francis' 2015 statement in the first ever papal encyclical on the environment: "God always forgives, man often forgives, but nature never forgives." I think he is speaking to nature's impersonal honesty. Panarchy works so well because it knows only the truth. Gravity isn't going to lie or offer compassion when you drop a rock on your foot. The laws of physics show no favoritism. When the arctic warms, ice melts. When ice melts, sea-level rises. Too few calories and organisms die. When a helmsman disregards depth charts, the ship runs aground. Self-aggrandizement by governors and brain washing of the governed corrupts feedback. When falsehood silences or corrupts guidance there is always a price to pay because nature knows no grace. The future hangs heavily on our ability to learn and change.

Reflection on Quantum Mind

I am not a physicist and do not pretend to understand the finer points of quantum theory. It suggests a whole range of doors which we have yet to open. John Archibald Wheeler often quipped that man's limited knowledge was like an island in the midst of a vast ocean of ignorance. As our knowledge increases, so also does the shoreline of our ignorance. Paradoxically, the more we learn, the less we seem to know. Quantum theory creates a boatload of new questions.

I am curious as to how much of the memory of the universe is encoded in the quantum wave function of the elements that comprise our bodies? Since our bodies are made of recycled stardust and incarnate starlight, does our every cell hold the memories of supernovas and nebulae and the gradual accretion of our solar system by gravity?

And the neurons of our brain which are logically derived from that entire series of prototypical iterations housed in millions of living species—do they remember, somewhere within the inner sanctum of the coherent wave functions of each neuron, what it was like to be a tree, a bacterium, a slug, a reptile, a bird, Homo erectus two million years ago, a Neanderthal, or a Clovis person?

Carl Jung would have had a field day with this. Jung, as many readers know, said that the "collective unconsciousness" is essentially nature within us as a whole passel of primordial memories from the various states of our evolution which are inaccessible to the conscious mind.

In later life, Jung worked with Wolfgang Pauli to develop his theory of *Synchronicity*. He appears to have tried to link the concept of *quantum entanglement* with subconsciousness to explain how events distant in time and place might affect the state of mind. Einstein famously called entanglement, "Spooky action at a distance." It appears that entanglement does have some reasonably important effects, perhaps such as photosynthesis without which life could be problematic. But linking disparate happenings in time and place with changes in one's mind today isn't very widely accepted. Most physicists and psychologists consign it to the realm of pseudoscience—the bailiwick of cranks and charlatans.

And yet, I cannot get over the sense of remembering, when a full moon rides up over the marsh on a benign night in spring, enchanting the world, that I have been in this place and witnessed this event a million times. It goes deep into my being. I identify with the peepers' Hallelujah Chorus and the moon's silvered beams glinting off newly formed sedges and waters just released from winter's icy shackles. I know this place. I am this place and this light and these songs at the beginning of creation once again.

Is this the sensed completeness of pieces of Cosmic Mind reconnecting with itself?

Mysticism? Pseudoscience? Perhaps yes, and perhaps no.

5. Causality

"Everything happens for a reason."
— Folk Wisdom

"There are no structures that are not linguistic ones ... and objects themselves only have structure in that they conduct a silent discourse, which is the language of signs."
— Giles Deleuze[1]

"Life in this picture is all one dynamic system of information processing in which every level of organization is both causing and being caused by other levels."
— (as described in Farnsworth et al., 2013)[2]

Information and Causality

WITH EACH PASSING DAY, winter's ice retreats further from the lakeshore. The fan of 20-Mile Brook is now open and eating its way slowly toward the center of the bay. Wood ducks, mallards, and great blue herons gather there for nourishment and shelter, dipping, preening, and swimming in circles until surrounding ponds have broken free from their icy shackles.

Locals have a betting pool to see who can guess, to the closest day and minute, the time when the Big Lake (Winnipesaukee) will be ice-free. I'm estimating sometime during the last week of April. But it is difficult to pin down accurately since causality is random and chaotic. Two things factor in: first, the amount of sunshine and temperature, and second, wind direction and speed. Ice prophets do about as well as stock pickers. Causality is real, but it often refuses to march to our expectations in a world where small changes result in big differences.

Millions of tons of ice will soon disappear, thousands of lives will alter course, events will happen next month that could not have happened last month. And all of this physical change is predicated on immaterial photons. The logical pattern of information in water makes life possible on Earth.

Like our mental exercise to establish causality for ice-out, philosophers have been reverse engineering phenomena for thousands of years to determine original cause—Ultimate Reality, God, Primal Cause, Logical Cosmic Principle. Earlier on, the world was understood as material alone. It was a simpler place where observation and reality were one—the sun rose out of one ocean and descended into another for a period of rest. The world was flat.

According to the aboriginal mythology of the Hopi people in the American southwest, Spider Woman (a more pleasing account than Ovid's Arachne) weaves the web of the universe. After that, she creates clay figures in her underworld chamber and sets them aside. Soon, however, she notices that they are moving and have life. Understanding their need for light and continuance, she creates a hole in the earth, a sipapu, from which they emerge as lizards that soon metamorphose into humans. Even today, a hole in the Earth exists in the center of Hopi kivas to remind them of their origin.

Every culture has its own genesis myth. For life to have meaning, we need to have a story to help us understand the cause behind life.

Aristotle used the word "intelligible," but it means the same thing. We are driven to explain things by cause so we can understand. Once we feel we understand, the pieces fit and meaning emerges.

Our grandchildren love puzzles, so we often take a new puzzle when we visit. This gives occasion to sit together and catch up on events while engaged in mental exercise. Initially, as the unconnected pieces are dumped on the table it's just a meaningless jumble. The challenge is to derive meaning from randomness by a process of reverse engineering. We select the four corner pieces, then all the edge pieces, and begin putting them together based on the logic of their color match and the reciprocal fit of each particular tab and blank. Working backward like this we eventually arrive at the full meaning of the puzzle after each piece is joined in its one and only specified place. In deciding placement, information for shape and color match are the causal agents. But they were selected or rejected by the logic of color, pattern, tab and blank, just as every system and living thing was selected by thermodynamic basins of attraction to fit in the grand tapestry of life and Earth.

Deutsch and Marletto, physicists at Oxford University, have proposed the *Constructor Theory of Information* as an attempt to unite the concepts of classical information with information contained in the quantum wave function. Construction takes place during the interactions of a two-part physical system. One part, information encoded in an unchanging systemic pattern called the *constructor*, is causal for transformations, yet remains unchanged itself, like genes in biological systems or memes in cultural systems. The other parts are physical substrates which are rearranged into patterns defined by coded information to exhibit new emergent properties and functions.[3]

Evolution is the continual refinement of algorithms for fit within dynamic basins of attraction. Continual algorithmic recombination resulting from genetic mixing during reproduction, mutations, and horizontal gene transfer between species, offers a perpetual stream of new possibilities that are selected by niche affordances for continuance and reproduction. It's the universe's way of saying, "May the best algorithm win." Information constructs. Nature, via basins of attraction and the laws of thermodynamics, selects.

These concepts are not easily grasped at the outset. We find ourselves back with Plato in the cave, confident then confused about what is real. We are drawn to things we can see because the brain likes visual candy. Information and algorithms, while understandable, are less accessible, colder to the brain. This has been a great source of cultural conflict ever since Darwin. It is easier to identify with Spider Woman than with impersonal and immaterial evolution that runs on logical informational rearrangements.

Choice

The power of the digital age rests on information reduced to binary digits—0s and 1s. As such, response to difference is answered as yes or no, both or neither. Informational "about-ness" is its probable value. If the temperature is 15°F outside, that information has a direct "about-ness" for the way we dress, whether we walk to school or ride, bring the dog into the basement, or check the level of heating oil in the tank. We give relative weight to information and that weight has causal value. A wind speed of 10 mph outside has low weight compared with a wind speed of 60 mph. Value decision will assign 60 mph wind speed a much higher probability for negative outcomes than 10 mph. When logic no longer squares with informational reality, society enters Kuhn's crisis territory. The Greek word *krisis* means a "parting" or "estrangement." Divergent stories of how the world works and came to be are causal for the crisis in modern society when logic is perverted or denied irrationally. The absurdity of anthropogenic climate change denial is just one manifestation of a crisis between divergent stories.

Our life direction is the result of multiple choices. We are where we are today because of choices we made in the past. Some were logical and beneficial. Some were illogical and left scars. Life is information processing where choice in one system impacts other systems or functions based on shared feedback. The overall well-being of the organism is predicated on these choices, most of which are subconscious.

Information as Signal

A dematerialized science has moved the basis for intelligibility from matter to unseen abstractions that move and pattern the physical world. These are the new structural sciences such as cybernetics, systems theory, complexity theory, information theory, and semiotics. They are called "structural" because they are the causes that organize mind and matter.

This implies that the structural sciences can serve as a bridge between nature and society. The structure of information becomes a bridge to cross the chasm between the world of nature—oxygen, lungs, blood, muscle tissue—and the world of mind, life, and meaning that makes continuance and culture possible. Informational structures are the causal relations that unite the sciences and humanities—evolution with culture, biology and healing, systems theory with law, genetics with psychology, pattern with religion, and ecology with economics. This is no small matter. Without this bridge life cannot be understood.

Biosemiotics is one of these new structural sciences. It studies the signal/response mechanism in living systems. As a little boy I watched the TV series featuring the Lone Ranger and Tonto. A few of the programs involved the use of smoke signals as communication between Native Americans. Smoke was the physical agent that carried immaterial information (something about something) that was causal for human decision.

These are linguistic structures just as spoken language is an arrangement of signals designed to transfer information for the purpose of knowing something about something. Language is relational and those relationships issue in meaning. Choice is predicated on signs (messages). Signals between the environment and life systems are causal for response.

Imagine, you are camping in a wilderness area that is home to grizzly bears. You awake in your tent at 3 AM to the sound of heavy steps in the woods near the tent. Instantly, you remember the food bag near your feet... Or, you are sitting outside your home, watching a fire in the fire pit while waiting for friends. You hear heavy footsteps... The signal in both cases is identical but your emotional reaction is at opposite ends of the spectrum and is determined by linguistics—by the arrangement of information relative to other information. In the first instance, location, vulnerability, time of night, habitat of grizzlies, and food in the tent combine in logical syntax to set off the semantic alarms by the release of adrenaline—I might become bear food. In the second scenario, the sound of heavy footsteps may cause the release of endorphins because you associate the signal with expected good times.

Signals are the immaterial bond between the organism and its environment that determine behavior and response. Every day we receive, sort, analyze, and respond to millions of bits of information to read the logic of patterns that determine their value to our endeavor. Language is a natural outcome of all forms of interaction. Earth and sky, sea and sun, rain and land, man and dog, wolf and moose, bacteria and virus—each uses a language older than words.

Thus, language is a property of Earth. Each lifeform evolved to occupy a particular thermodynamic niche where it could obtain energy to survive and pass its coded information on to the next generation. Life altered niches and niches altered life by signaling. Today's life forms could not exist in the original Earth environment, heavy with carbon dioxide, hydrogen sulfide, and methane. But life changed the environment with greater oxygenation by cyanobacteria.

Every mammalian body is a cacophony of chemical signals. Peptides and neurotransmitters, gastric juices and foods, kidneys and salts, brain and blood pressure—each is in constant communication by the agency of chemical messengers flowing through the circulatory system. Consider the balance between bobcats and snowshoe hares around this marsh.

Hares are browsers of bark and herbaceous materials. If conditions are right, they reproduce like hares with up to four litters/year and four leverets in each. Yet, as their population numbers increase, hunting gets better for the bobcats and they produce more offspring. A greater number of bobcats consume more hares. As stress from predator pressure increases, hares secrete cortisol (a stress hormone) that decreases fertility, causing females to have fewer and smaller litters. Higher cortisol levels are actually passed onto offspring causing the next generation to have fewer and smaller litters. The result is a hare population crash, which induces a bobcat population crash. In this subsequent environment with lower stress, hares begin to reproduce at more normal rates, causing the population to increase which causes the bobcat population to increase. It's the old boom and bust cycle that has played itself out for millions of years. And it's all caused by a language of signaling—a logical pattern of information and meaning.

Earth is one great communion. Crustal plates in-form mountains. Air currents speak in the words of weather. Ants and bees signal by a language of dance, pheromones, and timing. The quantum wave function is a volume of facts at a multitude of observation points.

Jesper Hoffmeyer describes signaling between a species of squid and bacteria in his *Biosemiotics: An Examination into the Signs of Life and the Life of Signs* (2008).

Hunting small fish off the coast of Hawaii by moonlight, a species of squid becomes visible to predators below by its moon-cast shadow. Evolution has selected a symbiotic relationship between the squid and a bioluminescent bacteria that live within the squid's mantle cavity. Because the bacteria emit light in the range of moonlight, predators swimming below cannot see the squid above. At daybreak the squid buries itself in the sand and expels up to 95% of the bacteria. Because these lower bacterial concentrations are beneath the quorum threshold for transcription of the lux operon genes, the squid does not glow and become vulnerable in the sand. Logic manifest in systemic feedback between organism and environment selects adaptation for continuance.[4]

A clear signal integrated into systemic operation via feedback fosters growth and continuance. Periodically, however, the signal is either corrupted or interrupted. Faulty signaling in living systems is comparable to malware in computer systems which corrupts programs to yield faulty outcomes, become illogical, or crash.

Many researchers are framing cancer as an information-theoretical problem where cellular feedback logic has been compromised. In this case, cancer becomes an infinite loop with no stop mechanism, thus beyond normal methods of control. Cancer cells may produce camouflage proteins that fool T and B lymphocytes into thinking that they belong. Otherwise, the lymphocytes would destroy the errant cancer cell. In other cases, epigenetic factors (stress, chemical pollutants, poor diet) contribute to the methylation of DNA bases and suppress certain gene transcriptions which would have defeated cancer cells. The immune function is based on recognition of antigen patterns, communication between system and lymphocytes, and recognition of foreign cells based on stored memory. When these abilities are compromised, cancer often gets a pass that allows it to run free. Corrupted signaling is causal for immunological breakdown.[5] Disease is a general failure of systemic logic most often traceable to corrupted/suppressed information. A twisted understanding of free speech that permits deliberate lies to run rampant in public media may have the same outcome in our social/political system.

Epigenetics

Over the last 50 years scientists have made huge strides in linking disease, psychological manifestations, and social discontent to environmental factors. This is a logical outcome of greater ecological understanding, for man is a creature in communication with his environment. We affect the environment and the environment affects us, even to the level of genetic expression.

Geneticists understand that genes are "plastic" and can express in different ways under different conditions. This has developed into a whole new field of study known as "Epigenetics."[6] The prefix "epi" means on or over. Although geneticists understand that a certain gene governs the manifestation of a certain outcome, they have learned that the outcome is not fixed in stone. Environmental factors such as temperature, psychological stress, chemical pollutants, different enzymes, etc., can affect gene transcription and expression.

Life is a two-way communication using the language of signals that affect causality in organisms and the Earth. We react to toxins in the environment through sickness, stress, and lost productivity. Earth is reacting to human activity by species extinctions, altered weather patterns, rising sea levels, and changing agricultural productivity. The conversation is becoming increasingly antagonistic because signals from our stories (religious, political, cultural, and economic) have caused us to disregard honest environmental feedback and make course corrections.

Language lives between two or more parties—sender and receiver—with roles continually oscillating. The Earth is one party of this conversation and governs the syntax within which we were formed. Meta-Logic governs the syntax of Earth's messages. It's the world that Black Elk knows in John Neihardt's American classic, *Black Elk Speaks*. For Black Elk and his people, the Lakota Sioux, Earth is a sentient being forever speaking and instructing people in right relations and attitudes. Near the end of his life, Black Elk summarizes some of this in a conversation with Neihardt, "Everything the Power of the World does is done in a circle. The sky is round, and I have heard that the Earth is round like a ball, and so are all the stars. The wind, in its greatest power, whirls. Birds make their nests in circles, for theirs is the same religion as ours. The sun comes forth and goes down again in a circle ... Even the seasons form a great circle in their changing, and always come back again to where they were. The life of man is a circle from childhood to childhood ...."[7] For aboriginal cultures, Earth is always the dominant communicator. Man must accommodate Earth.

Western Culture has assumed a different point of view. Because of our cultural/religious narratives, we assume that man is both the pinnacle of creation and dominant force on Earth. The Earth is here for man to both use and govern. Fashioned in the image of God, man is to dominate Earth and bend her to human will. Our technological prowess has only enhanced this vision. In the process we, who are novices at management, are overruling systemic logic that took vast eons of time to build. Over this last century, Earth has begun to give negative feedback by decreased bio-systemic performance, life extinctions, weather patterns, disease, and social breakdown. In many cases, we have become tone-deaf to Earth's utterances, continuing on in the face of negative signals. Confused, corrupted, and stifled signaling, in non-linear systems, can have all sorts of amplified and unintended consequences.

The world of tomorrow will be built by the choices we make today.

6. Complexity and Chaos

"There is grandeur in this view of life, with its several powers, having been originally breathed into a few forms or into one; and that, whilst this planet has gone cycling on according to the fixed law of gravity, from so simple a beginning endless forms most beautiful and most wonderful have been, and are being evolved."
— Charles Darwin

"This interconnection of all created things to each other, brings it about that each simple substance is a perpetual, living mirror of the universe."
— Leibniz

"The system of the 'universe as a whole' is such that quite small errors in initial conditions can have an overwhelming effect at a later time. The displacement of a single electron by a billionth of a centimeter at one moment might make the difference between a man being killed by an avalanche a year later or escaping."
— Alan Turing[1]

Complexity

When I was a little boy, mother bought a small craft kit for me to weave potholders. It had a metal frame with hooks for attaching the warp and woof, plus a bag of premeasured and cushioned threads. Thus, I learned elementary "com-plication" = "to weave together."

Things that are intertwined, interrelated, and hold together based on shared relationship, are complex structures. The success of these complex structures depends on the integrity of each thread within the greater whole. Because everything is connected, one broken thread could unravel the structure.

Economic, ecological, biological, political, and religious systems, while appearing superficially diverse, are all com-plicated. All are complex systems composed of subsystems composed of groups composed of units, where each piece and part communicates via feedback to indicate its current state as it affects the function of the whole. One broken connection could affect the survival of the pattern. Cybernetics gave rise to the study of complexity.

The Santa Fe Institute began in New Mexico's high desert in 1984 when a group of diverse scientists and mathematicians first met to study the broad "complexities" of multi-thread systems. Today, the Santa Fe Institute is one of the world's premiere hubs for the study of complex systems. In her book *Complexity* (2009), Melanie Mitchell, an external professor at the Santa Fe Institute, lists the four areas of study that are basic to complex systems: "information, computation, dynamics and chaos, and evolution."[2] She defines a complex system as, "A system in which large networks of components with no central control and simple rules of operation give rise to complex collective behavior, sophisticated information processing, and adaptation via learning or evolution."[3]

Complexity theory is part of nonlinear dynamics, or nonlinear systems theory, or dynamic systems theory. Linear, or straight line, systems are more scalable—small inputs yield small outputs, large inputs yield large outputs. Newtonian physics is linear, more like clockwork where one less swing of the pendulum each minute causes the clock to lose one second/minute, or one minute/hour, etc. In the mathematics of complexity, small changes in input can have exponential implications for outcomes. Therefore, it is nonlinear; it is non-proportional.

Complexity is all about patterns and the relationships within those patterns. Things are bound together in a fabric or map of connections that are influential for computed outcomes. For instance, if, at the beginning of a journey, one turns onto the wrong road, that change could result in missing the desired destination by hundreds of miles at the end of the trip. A change of sign (+ or -) is like choosing to drive east or west. Small initial changes result in huge differences of outcomes.

Consider the famous *Butterfly Effect* used in meteorology to posit that a butterfly flapping its wings in China might tip the balance of air currents to cause a tornado over Oklahoma. Like algebra, small initial changes in variables—small perturbations—are amplified and multiplied by self-reinforcing feedback and could produce disastrous outcomes. This part of complexity is often referred to as *Chaos Theory*. Readers might enjoy Cixin Liu's novel *The Three Body Problem*, a fictionalized account of Henri Poincare's three-body problem where even the smallest variations in inputs yield massive and unpredictable results.

Bacteria are primitive forms of life. And yet, even a single bacterium contains around ten million bits of genetic code. Changes in just a few bits of that code may be the difference between benign and infectious. This is not idle speculation because hundreds of species of bacteria live in harmonious relationships within the human gut, digesting foods, secreting enzymes, affecting nutrient balances, even aiding our immune system. Multiply out the possible scenarios for change by altering just a handful of genes or their protein expressions in our small intestine. Any number of changes could prove fatal. The amazing thing is that we are so healthy for so long, going years without a glitch and completely ignorant of the immense complexity we carry about at every moment in our enteric cavity.

Harvard sociobiologist E. O. Wilson has spent nearly a lifetime studying ants and ant colonies. He uses the term *super-organism* to refer to an ant colony.[4] A single ant by itself is reasonably complex. Yet, when combined in a colony, a multitude of new complexities emerge— division of labor, elaborate tunneling and chambers dedicated to different functions, the use of timing and body movements to convey messages, pheromone trails laid in different concentrations to indicate the pathway to and amount of available food. Although the queen is responsible for laying eggs, she has no governing control over the different activities and functions of the colony. Decisions are made by quorum among those workers contributing to that function. Determination of sex and class (worker, guard, male, female, nurse) is done by the nurses. Storage of food and cultivation of symbiotic fungi on that food is done by workers. (Ants were farming for hundreds of thousands of years before man.) An ant colony is panarchist, i.e., government and decision making is an emergent and holistic process that must be ascribed to the entire organism, not to any central authority. Each unit functions for the best long-term interests of the entire organism. Integrity and continuance hinge on good connections.

Like an ant, a single isolated neuron isn't good for much. Mind is an emergent quality based on the networking of multiple neurons that form patterns to govern various functions within the body and brain. What matters most is the logic of the patterns that determines how they will process information for holistically positive outcomes. Ants secrete pheromones; neurons' secrete neurotransmitters. In both cases, information is relayed via chemical messengers. Ant colonies operate on a relatively simple number of causal patterns. If Ray Kurzweil is correct, the human brain is continually referencing a library of 300,000,000 patterns to weigh current environmental affordances and compute for optimal outcomes. Add to this the tens of thousands of different proteins and receptors involved in every thought process and imagine that just one protein is printed with a deficient methyl group. Because of this minuscule deficiency an entire process could break down, just as a few alterations can cause the build-up of plaques that are symptomatic of Alzheimer's disease. Claude Shannon, one of the most brilliant men of the 20th century and leader of the information age, died of Alzheimer's, completely oblivious to his role in changing the world. Broken connections can lead to massive changes in complex systems.

Since the human body is an open system, all sorts of pathogens have multiple points of entry. Cold viruses swirl in the air after an infected person coughs. HIV enters the body via the orifices. A misstep onto a rusty nail opens our body to tetanus. In nearly every case, antigens (infectious agents) attempt to commandeer the machinery of human cells to reproduce more antigens. In order to do this, however, an infectious agent must evade, escape, or confuse our immune system.

The genius of our immune system is its ability to "read" the proteins of antigens for identifying code, like the barcode scanner at the grocery store. Lymphocytes have the amazing ability to learn, remember, and transfer patterns of both good and malign cells, to attack invaders while passing by the multiple thousands of proteins, chemicals, good bacteria, and cells necessary for the integrated function of our bodies and their processes. When a novel antigen is detected, it will be tagged by a B-cell for destruction. But since it is new, not all lymphocytes may recognize it as an invader. But as B-cells reproduce, memory is transferred to daughter cells so that more and more of them recognize the invader and mount an ever-stronger defense with each new generation. This process mirrors Darwinian natural selection. It is mind sensing, acquiring, and passing on new learning to succeeding generations. In this manner, people infected with Covid-19 will, if they survive, develop some form of immunity. Defeat of malware issues in restored connections which we define as health.

Cumulative modularity is a common trait running through all complex systems. Consider the word processor Microsoft Word, now in its 14th generation. Beginning with the first generation, coders who had worked for Xerox developed the platform for Word. As inadequacies and errors appeared, code was corrected and added to produce new emergent properties. Modules from each generation were cut, pasted, improved, tweaked, and combined to form an ever more complex program with new capabilities. Somewhere in each iteration, sections of code appear that are common to all, but placed in novel combinations with newer sections of code.

The DNA code to make a human has advanced in similar fashion. Some of our most important strings of code go all the way back to bacteria. Source code for our human mitochondria was traced by the Carl Woese team at the University of Illinois (1985) to the protobacterium Agrobacterium *tumefaciens,* the source of galls in walnut trees and grape vines.[5] The little powerhouses that deliver the energy that keeps us going day and night originated with a pesky species of bacteria, and have been handed down across multiple genera of organisms until they became human.

The next time you bait a mouse trap with peanut butter and catch the mouse that's been nibbling your cheese, give it a good look-over. Then consider that 95% of your genome is common to that of the little mouse in your trap. Or the enzyme that metabolizes glutamine is identical in you and in the trees now bending before an oncoming storm. Or that we are 98% chimp genome and carry from 1%-4% of the Neanderthal genome. Your genes once swung from trees in an equatorial jungle and used red ochre to paint bison on cave walls 30,000 years ago. Complexity arises from snipping, pasting, adding, refining, and mixing in novel combinations that yield all manner of unexpected traits and qualities. We are all—mice, bacteria, and Neanderthals— one great interconnected continuum.

Let's go for a walk in the woods. The scene is relaxing. Our surroundings are soft and green. Gentle cumulus clouds float overhead. A fragrant scent of balsam and fern and drying earth wafts about our nostrils. Mushrooms hold aloft their little domes and moss softens glacier-pummeled granite. Do you find yourself at ease? Do you sense connection, a feeling that you know this place, that you are among friends? In short, do you sense the inklings of identification? Fine, now let me make a suggestion—perhaps you identify because you are among family. Remember the previous example of a word processing package and how each new generation was simply a recombination of old code plus recent novelties? Isn't it the same out here in nature? When you look at a fern, an oak tree, the chickadee that almost landed on your shoulder, the salamander under a log, leaves that are slowly turning to compost ... perhaps you are seeing prototypical iterations of yourself? Their code is your code, sliced, recombined, and complexified. Perhaps your conscious mind never framed it that way, but your subconscious mind knows the truth and is saying as much through your feelings of belonging and identification. Is that what Leibniz meant, "...that each simple substance is a perpetual, living mirror of the universe"? Or William Blake's allusion, "To see a world in a grain of sand, And a heaven in a wildflower..."?

It appears that ever increasing complexity is the nature of Nature.

Chaos

The original Greek, *xaos*, implies emptiness, or an abyss from which indeterminate potential may arise. This is the "void" of Genesis 1.

Three major paradigm shifts have occurred within physics over the last one hundred years—relativity, quantum mechanics, and chaos. Each has been an assault on reductionist science and the orderliness of Newtonian mechanics. For more than two thousand years, science sought to understand the world by understanding its constituent parts. The various scientific disciplines operated largely as stand-alone fortresses holding sway over biology, chemistry, geology, physiology, physics, etc. Few saw the need for cross-disciplinary communication and cooperation. Cybernetics, systems theory, and chaos have upended this pattern.

The concepts of non-linearity, circular causality, indeterminism, and semiotics as command and control gradually replaced the old patterns of understanding. In the 1960s, theoretical meteorologist Edward

Lorenz was using a computer to find a better model for weather forecasting. Rounding his numbers to only three decimal points, Lorenz was able to repeatedly replicate certain models. When another decimal point or two—exceedingly small differences—were entered into the same model the results were drastically different. Small initial inputs produce exponentially larger changes in non-linear systems. The results were chaotic. Edward Lorenz was the first to propose the *Butterfly Effect* mentioned earlier.

Linear systems are more additive; circular causality generates feedback that can multiply outcomes geometrically. Systemic illogic (malware, non-truth) is amplified by circular feedback over time. Built-in systemic resiliencies may mitigate falsehood for a time, but beyond tipping points, however, mitigation efforts are useless. Disintegrated systemic parts eventually settle into lower basins of attraction with fewer emergent qualities and networked abilities.

A young mathematical physicist named Mitchell Feigenbaum (1944-2019) came to Los Alamos Laboratories in the 1970s, thinking more about clouds than problems he was hired to solve in fluid dynamics. With a wild shock of black hair, he looked like a cross between Einstein and Schopenhauer. It was here that he formulated his ideas about chaotic systems—period doubling bifurcations, universality, logistic mapping, and the Feigenbaum constant of 4.6692 that plots period doubling bifurcations in all manner of nonlinear systems. Because the mathematics of chaotic systems is universal, it united a whole cadre of cross-disciplinary scientists and mathematicians to look at the bigger picture of interconnected patterns whose pieces were formerly studied in isolation. Henceforth, it would be impossible for the physicist to neglect biology and psychology, history and philosophy, cybernetics and economics.

Chaos refers to a condition of apparent disorder and unpredictability. Fractal geometry—the analysis of uneven and irregular surfaces and edges—is a part of chaos studies. Shorelines and cloud edges often yield patterns and symmetry unimagined at first glance.

Meteorology is the most fertile field for the study of chaos because of the unlimited and varying inputs to weather—cloud cover, precipitation, fluctuation of the jet stream, sunspots, latitude, pressure systems, solar absorption/reflection, wind speeds and direction, etc. Even with today's supercomputers, weather forecasting is never definite and generally expressed in percentage probabilities. Slight variations across a whole range of inputs can dramatically change manifestations. And that's the best-case scenario when we are dealing with known factors.

Then there are all the "unknown unknowns." What if a coal-fired electrical generation plant on the Ohio River suddenly fires up to meet increased demand? All the heat, particulate matter, and steam released into the atmosphere will affect weather downwind for hundreds, if not thousands, of miles. Perhaps a farmer in Kansas is burning brush to clear a new field? If the snow lasts longer in Illinois it may affect winds in New York? It is impossible to build unknowns into a predictive apparatus. Future probabilities for stock prices cannot factor in all the unknowns, like a coronavirus, whose slightest genetic variation produced unimagined consequences in global financial markets.

We walk a tight rope between prosperity and dissipation, never knowing whether the slightest wind will push us from one into the other, if a crashing wave will bring food or destruction, if an election will result in democracy or fascism.

The Methane Bomb

One previous unknown that is becoming more known is called *The Methane Bomb*. Methane is more generally known as the natural gas (not propane) millions of people use daily for cooking and space heating. Methane is a natural product of the decay of organic matter by bacteria. Cow burps and human farts are a product of gut bacteria digesting foods. All organic gardeners know that the compost pile gives off heat, even on frigid winter days, because bacteria are metabolizing organic materials into compost plus heat plus carbon dioxide and methane. Heat generated within the compost pile actually accelerates decomposition. Composting organic material sometimes generates so much heat that piles will spontaneously combust by positive reinforcement. This is common in large, old sawdust piles.

Twenty percent of the landmass in the northern hemisphere is located in subarctic regions known as tundra and boreal forest (taiga). With average annual temperatures below freezing, soils remain frozen year-round in a condition known as "permafrost." Normally, organic matter that falls to the ground in warm climates decays in short order. However, in frigid regions bacteria are inhibited and cannot digest organic matter. Therefore, organic matter accumulates in layers, is compressed, and becomes buried under newer layers. For this reason, these subarctic regions are a treasure trove for fossil hunters. Intact mammoths have been excavated from tundra permafrost, some with chewed grasses still in their mouths.

For over 10,000 years, organic matter has been accumulating without decay in layers up to hundreds of feet in thickness. These permafrost layers contain 2X the carbon currently in Earth's atmosphere, which is more carbon than man has released since the beginning of the Industrial Revolution. Since the Arctic region is warming twice as fast as the rest of the globe, scientists refer to all this unreleased carbon as a ticking time bomb, hence *The Methane Bomb*.

NASA's *Global Climate Change* (Aug. 24, 2018) documented a new and serious concern not currently in calculations for future atmospheric warming. *Thermokarst lakes* are forming in Alaska where permafrost is thawing and sinking, permitting water to accumulate in newly formed depressions. Accumulating water sinks into the ground accelerating further thawing which warms the soil, which creates even larger ponds. Warmer soils allow bacteria that have been inactive in the permafrost to become active again. Bacteria begin doing what bacteria have always done—digesting organic matter to release carbon dioxide, methane, and heat. Released heat positively reinforces more thawing to accelerate the process exponentially. Current mathematical models indicate high probabilities for a tipping point for massive methane release by 2040. Each of the above factors will positively reinforce others causing methane release to increase exponentially. A system once benign will flip to become an out of control dynamic accelerating global warming. Unknowns will then become knowns.

The release of methane will not be a linear process, but a circularly causal dynamic self-influenced by systemic feedback. The tipping point for accelerated release of permafrost methane, when it occurs, will be like a detonated "bomb," breaking innumerable connections between Earth and life forms that may issue massive die-offs.

In May 2013, the Mauna Loa Observatory in Hawaii measured atmospheric carbon dioxide at 400 ppm. The latest figure for 2019 is 415 ppm, or an increase of 15 ppm in six years. But the graph is not linear because the increase is accelerating. Most observers predict that by the year 2050, with business as usual, concentrations will be at 500 ppm. This could result in an average global temperature increase of 5° C, or more, by the end of the century. At that point the Amazonian jungle would become desert, agriculture in the subtropics would become unproductive, human migration to cooler areas would exacerbate political tensions ... and that's only the beginning.

Consensus opinion within the scientific community is that we are facing a disaster the likes of which humanity has never seen. As the arctic warms, ice and snow melt. This decreased reflectivity will cause land and darker sea to absorb more heat than before. Warming air melts more ice and faster. Greenland's glacial melt will accelerate sea level rise. The Antarctic ice sheet will follow suit, with more sea level rise. A warmer and dryer boreal forest is already under attack from bark beetles which are killing spruce and fir trees by the hundreds of square miles. Vast fires are consuming forests turned to kindling, leaving the earth black and burnt. A blackened surface without the shade of trees will absorb even more heat, faster, melting permafrost at an accelerating rate. Methane is 25X more potent as a greenhouse gas than carbon dioxide. In short, greenhouse gas release in polar regions will be like the proverbial snowball rolling downhill, continually getting larger and rolling faster. The world of 2050 may be nearly unrecognizable when compared with 2020. The current compounding of industrial technologies driven by our cultural/religious/political/economic stories is leading the world to a place where logical informational patterns, built over hundreds of millions of years of adaptation and selection, are beginning to depattern—to dis-integrate—into a chaotic soup of pieces and parts. That chaos will be a direct result of the false stories, denials, rationalizations, and lies we continue to tell ourselves. The crisis of global warming is simply a manifestation of our own cultural malware corrupting nature's algorithms.

Resilience

Generally speaking, all systems attempt to hover at or near homeostasis—a state of balance. Balance is maintained by the cybernetic process of feedback. Natural systems are panarchist, having no central decision-making authority (like ant colonies and the immune system), and depend on honest feedback to control the system, much like a governor maintains the speed of an engine. Adding new features, overriding limits, discounting feedback, and corrupting signaling compromises a system's internal and distributed governance and introduces chaos. Left unaddressed, systems in a chaotic state overshoot traditional boundaries. Patterned connections begin to break as the system destabilizes. The algorithm becomes absurd, incapable of managing conflicting information. It is dumbed down.

Rising global ecosystemic stress and deterioration has led to an increase in resilience thinking and planning.[6] Resilience is a type of systemic plasticity—the ability to absorb shock and spring back into shape. The cornea of the eye is resilient. When scratched, it generally repairs itself within a day. A surgical incision, when properly protected, will heal itself in short order. Hardwood trees store sap in their roots, drop leaves, and hunker down before the advance of winter. In spring they bounce back and begin a new cycle of growth. All life systems have built-in resilience because environmental conditions over billions of years have fluctuated within certain parameters. The bears that roam the forest around my cabin could never get enough calories to survive this last winter. Natural selection built a resilience mechanism into them in the form of decreased metabolism for a long period of time. Having stored enough fat in the fall, they are able to reduce caloric needs during hibernation and emerge again in spring when foods are more plentiful.

The ultimate non-linear outcome, however, is that the Manichaean Devil's bifurcations can become the source of Logic's creativity. Using his constant of 4.6692, Mitchell Feigenbaum plotted accelerating bifurcations at points 4, 8, 16, 32, etc. as systems erupted into chaos. Amazingly, however, stable patterns ultimately returned. Under the heap of disorder exists an unimagined order beyond obvious reason. He was seeing the same thing that Benoit Mandelbrot (1924-2010) saw in his new mathematics of Fractal Geometry. The morphogenesis of branching dendrites and branching trees and branching ice ferns and economic systems all work according to fractal scaling.[7]

Kayaking back to the shore of 20-Mile Bay last December, I witnessed a new order arising amid thermodynamic disintegration. It had been cold for several weeks, with steam rising from the lake each morning at sunrise. Knowing my days on the water were limited, I had paddled several five- to eight-mile circuits during the preceding weeks. This afternoon, however, I had gone farther than normal and was coming back to the beach after sunset. Skies were clear, stars had emerged, and the thermometer had plummeted. About a mile from the beach, the kayak began to crunch through thin layers of forming ice. Stopping to see what was going on, I sat quietly and actually saw ice crystals, etched by latent light against inky black waters, lengthen like silver daggers and expand into crystalline patterns. One ice dagger would unite with another and another as water went through rapid phase transition from liquid to solid. The entire surface system of the lake was bifurcating due to the loss of thermal energy, flipping within minutes from water to ice with an entire set of emergent properties. Had I remained much longer the kayak would have become like Shackleton's *Endurance* in Antarctica. Out of the disorder of increasing entropy, a beautiful new order of fractal patterning had emerged ...

Rapid and massive investments in renewable energies, electrified transportation (including high speed rail), green building codes, heat pump technologies, lower meat consumption, public policy to encourage lower birth rates, the education of women in developing nations, shortening supply chains, less air travel, and efforts toward a zero-throughput economy would quickly reduce our annual carbon output. Perhaps political and economic chaos will lead us to reorganize as a decarbonized economy. Or nature will deselect the agency causal for broken connections and move on without us.

Transferring subsidies from fossil fuels to renewable energy will yield outsized returns in the future. Health care costs would decline since fossil fuels contribute to all sorts of respiratory ailments, heart disease, and cancers. Green manufacturing will boost our domestic manufacturing base. The worst-case scenario, even if global warming were a hoax, is that we would have cleaner air and water, fewer wastes in landfills, a healthier population, fewer extinctions, better jobs, and a more vibrant economy. In the end, the chaos sown by campaigns of denial and disinformation may be the very agent to usher in a new energy and economic narrative.

7. Life

"We may have begun to understand evolution as
the marriage of selection and self-organization."
— Stuart Kauffman[1]
"Nature is genesis."— Holmes Rolston, III[2]
"The working of mind is life."
— Aristotle
"Life as, 'information that copies itself.'"
— Adami and LaBar[3]

What Is Life?

"THE CONDITION THAT DISTINGUISHES animals and plants from inorganic matter, including capacity for growth, reproduction, functional activity, and continual change before death." — Online Dictionary

While this definition sounds logical, not everyone agrees. A recent article in the BBC (Jan. 2017) states that there are 100 definitions of life. Physicists, chemists, computer programmers, virologists, geneticists, and biologists apparently have different points of view.

At the far end of the spectrum might be the artificial intelligence researchers who are conceiving non-organic future life powered not by metabolism but by electricity generated from solar panels. Our association of life with a vegetable garden might be meaningless to future inorganic robots. A field of grazing cows will become a vestigial carry over from evolution's archives. Who needs milk and meat when one runs directly on sunlight? Our electric car has no need of organic fuels, but it is still a car.

An information theorist, on the other hand, would tend to look at networked relationships and the feedback mechanism to describe life as the immaterial quality of mental process responding to information. Metabolism, reproduction, and growth are effects of life, not life itself.

Life cannot exist in isolation and is always in partnership with its environment. Life and the environment carry on a perpetual and dynamic reciprocity through the exchange of information, energy, and materials. In the process, life changes its environment and the environment selects for adaptive changes in life. One continually molds and reflects the other. In the early history of Earth, bacteria modified the environment in ways that furthered their own agenda, while the environment molded the coding of bacteria. The Earth and life transformed one another—they are a unit. Dualistic attempts to separate life from Earth are monstrous absurdities.

The genetic blueprint of every living thing is a plan for intended outcome. All future webs of relationships, structures, feedback mechanisms and systemic functions are contained in the initial coding—in the plan. But where did information come from in the first place?

Astrobiologist Paul Davies states, "Gravitationally induced instability is a source of information ... The ultimate source of biological information and order is gravitation."[4] It is gravity that compresses solar hydrogen into helium, plus the energy of fusion that is transmitted to Earth in the information of photons that run all of Earth's life systems.

Life Is Complicated

It's easy to regard prior iterations of ourselves as simpler forms of life. Yet, functions and processes are pretty much the same across the entire gamut of living things. Paul Davies states that the amino acids that combine to make proteins could be arranged in 10^{130} different combinations.[5] Others estimate up to a quadrillion existing proteins.[6] The human genome consists of about 24,000 genes. During sexual reproduction the genes from egg and sperm recombine into combinations never before seen on Earth. The statistically possible number of unique combinations of those 24,000 genes is nearly infinite.

Consider the human brain with its 100 billion neurons and each of those neurons having up to 10,000 connections. How could one ever draw a schematic diagram showing those connections? Such a drawing might begin looking like a spider's web. Then the lines would grow dense and close. In the end I think it would look like black ink spilled across a massive piece of paper. Neuroscientists offer the number of one quadrillion connections, but that is nearly meaningless because it is such an inconceivable number.

Sometimes when I walk in the woods, and conditions are right, I detect a slight odor in the air and know that a fox has passed recently, leaving his calling card on a nearby stump. How is it that within those 100 billion neurons and infinite connections with dendrites crossing dendrites that cross axons over axons, all bathed in a soup of ions and neurotransmitters, I instantly identify the odor as "fox" and not deer or garlic or decaying wood? Projecting into other lives, how is it that a polar bear can smell a seal from ten miles distant and know that it is a living or dead seal? Or how do Monarch butterflies that have a minuscule ganglia for a brain know how to navigate to a place in Mexico they have never seen? The complexities are simply unfathomable.

We are living in the ecological era, understanding more of the interconnectedness of life and the world every day. We are all in that same network. As Gregory Bateson stated over 50 years ago, "It is the pattern that connects." Whatever affects the pattern, affects all. We cannot understand life without understanding the complexities of the pattern.

Life Is Emergent

No person knows, beyond the shadow of a doubt, how life emerged. In the beginning Earth was a hot, chaotic, and lifeless place. Early on, the atmosphere was full of carbon dioxide with percentages of ammonia, methane, and hydrogen sulfide. We wouldn't have survived two minutes. But in time, bacteria appeared and began to use these gases and, in the process, changed the atmosphere to become a hospitable place for life. But where did the bacteria come from? That's the big question today as we search for the origin of life on Earth. Some think primitive nucleic acids could have arrived via an asteroid collision. Others look to thermal vents associated with volcanoes, both undersea and in places like Yellowstone National Park. Everything begins with that first replicator molecule. Was it a quirk of nature?

A chance combination of methane and ammonia in the presence of lightning? To date, there is no definitive answer.

Could life have emerged from this soup of Earth's early environment in obedience to simple mathematical principles that govern the way forces interact to attract and repel each other? Is it possible to visualize life as an emergent quality derived from the mathematical logic of interacting forces that settle into basins of attraction where they could extract energy and raw materials from the environment and replicate? If this is the case, then current life forms can be traced backward through their prototypical iterations to discover similarities within proteins, morphological structures, chemical systems, reproductive similarities, neural functions and anatomy, pattern equivalencies, and flow of algorithmic development (DNA and RNA).

Information is a difference that makes a difference. Any system that can make a yes-no choice based on difference is said to be mental. Science has demonstrated that natural forces, environmental affordances, and molecular electrostatic symmetries will repeatedly aggregate into logical forms. Chemists do it every day in the lab. It occurs day and night in the fractional distillation towers of oil refineries. Animals demonstrate the precision of immanent Logic governing algorithmic patterning every time a protein folds. **The working of mind is life.**

Goal of Life

Genes code for proteins which control traits, structures, and abilities. Therefore, genetic algorithms are teleological—they code for result. Logically patterned information implies a logically derived outcome.

Logically patterned information is intention. My drawings submitted to the building inspector were "prophetic" with regard to this cabin. They described ahead of time—before the first shovelful of dirt was moved—what he should expect to see upon completion. Plans are goal-oriented statements.

Therefore, transcendent, theistic creationism and built-in logicalism arrive at the same point. Both are directional and goal related. One says the multi-varied forms of life were stand-alone creations made to harmonize with Earth systems and continually replenish themselves into perpetuity. The other says that forms are logically determined, over long periods of time by the forces of nature, to fit within their environment for survival and reproduction. Both end with forms fit for specific environmental niches in order to survive and project their algorithms into the future. This is purpose or intention. The main difference is that theistic creationism places causality outside the system, separated to an inaccessible and unintelligible realm, while the other dictates that purpose resides within the system and is logically accessible, therefore intelligible.

This difference, however, is significant. An unintelligible process cannot yield meaningful information because we have no access. Information is light, and where there is no light one cannot see where he is going. Darkness isolates; isolation hides connections; and non-connection is non-relational. Since meaning derives from relationships, non-relational life cannot have meaning. On the other hand, immanent logicalism embeds us within the greater tapestry of time and the universe in an intelligible and meaningful way. We are connected by pattern and substance to everything that went before and that will come after. We are part of a continuum. We are integrated with every other particle of our universal family. We fit. We belong. We are at home in a monist universe.

The dualistic creationist paradigm that has dominated Western civilization for at least the last 5,000 years has left the West in an epistemological crisis and set the stage for the new story now being born. The great fruit of cybernetics, and the informational age that flowed naturally from it, is the understanding that no bit of information stands alone. Information is meaningful only in relation to other information.

The current western paradigm of dualistic creationism is epistemologically insubstantial and cannot bear the rigors of investigation. There is no substantive evidence for a transcendent personal God, heaven, hell, intermediate beings, or an instantaneous creation of living things unrelated to one another. As far as can be determined there is no self, no soul, no literal bodily resurrection of the dead or afterlife in paradise. These things are unintelligible because there is no demonstrable and repeatable evidence for any of it. We are trying to reach the bowl on the top shelf by standing on the hologram of a chair. Let me quickly note, however, that I am not rejecting these myths. They are freighted with great meaning but are metaphors of a higher order that are corrupted by literal interpretations.

If the goal of life is integrational harmony and continuance, **immanent logicalism binds us to the fabric of time and the world is such a way as to provide a sense of meaning, belonging, and fit.** If information is intentional (Latin: *to stretch toward*), there is a goal. The clear implication of bodily systemic regulation and a functional immune system is that life "cares" about life and continuance in order to meet that goal. The goal of life is the projection of information into

the future by way of reproduction. We do not project our "Self"—
that running narrative of which we are the main actor—but systemic
information in the form of our genetic code blended with that of
another individual into a unique form never seen in the universe.

Therefore, if nature and logic have a goal, we should be able to
have some idea by examining the history of life. Life began as
simple protobacteria, progressed slowly over time through ever more
complex forms all the way up to man. The main outcome has been
increasing complexity in all manner of novel forms. Every step of
evolution is marked by an increase in both the volume and complexity
of informational forms. Patterns become increasingly larger, with
more parts, systems, information storage, computation abilities,
responsiveness, and ultimately consciousness. Thus, if the observed
direction to date is any indication, one could speculate that the goal of
the universe is to unify the total informational content of the universe
into one integrated Monism or Singularity where all possible logical
patterns exist as a unity.

Life Is One

The Archimedes Palimpsest is a "Euchologion"—a Byzantine Prayer
Book probably composed in Jerusalem around 1229 AD. Under
examination by various scholars specializing in ancient documents,
however, the Euchologion revealed a deeper level of previously scraped
text. It turns out that the author of the Euchologion had no use for
Archimedes' mathematical principles and proofs, scraped them and
began composition of his prayer book. Archimedes (287-212 BC)
lived in Syracuse and spent time in Alexandria. We are most familiar

with his explanation of buoyancy, although that was just one of his mathematical efforts. Lacking funds for supplies, Picasso and van Gogh sometimes painted over formerly used canvases.

Perhaps we might use the palimpsest metaphor to understand the oneness of life.

Every living thing that has ever emerged on the face of this green Earth has a lineage that can be traced regressively all the way to the first bacteria. One might say that bacteria have been, "the bots of God." Bacteria are like the pieces of a Lego kit, and have been altered, mutated, recombined, agglomerated, complexified and logically patterned into all life forms.

Take a moment now and look at your house cat, if you have one. The same cat left a dead mouse at your door this morning. Look at the mouse, also. Fine, we look at a cat and a mouse and we see a cat and mouse, in the same way we looked at the Euchologion and saw a Prayer Book. But it was far more than that and carried deeper information that we could not see with the naked eye. I submit that the cat and mouse are biological palimpsests.

What you did not see when looking at the mouse is a creature that evolved over 11 million years ago. You saw, but perhaps did not connect, that your cat and the mouse are built on the same platform, with essentially identical systems built from mostly identical genes but expressed in novel ways for their adapted niches. Neural processes, digestive functions, circulation, immune system, reproduction, metabolism and structural form operate basically the same with the same molecules, cell structure and genetics in mouse, cat, and you.

I'm saying that if we scrape away the superficial ink and paint, underneath we are virtually the same information. Each lower form on the evolutionary ladder is simply a prototypical iteration of those that followed on, in fits and starts, with increased complexity. We are composed of the same units (proteins, cells, organs, systems) that have been aggregated and integrated with increasingly complex functions and abilities. At the same time, we have lost some functions. Arms aren't very good at flying and fingernails don't grip tree bark all that well. But a bigger brain has figured out how to do these things.

In the larger view, all of life is a unity with one iteration written over another all the way back to the blank parchment. We are systems built from and linked to one another. Consider the Greek roots of the word system, *syn* = together, and *istemi* = to stand. We all stand together.

We Contain Multitudes

"Do I contradict myself?
Very well, then I contradict myself.
I contain multitudes."
— Whitman

The old model for evolution, often visualized as a multi-branched tree, has begun to fracture. David Quammen's *The Tangled Tree* (2018), counters Darwin's vertical progression of life as depicted in the metaphor of a tree—a progressive sequence—with the image of a web of life, where causality is often horizontal and across species.[7]

Dr. Carl Woese (1928-2012) and his research on molecular phylogenetics at the University of Illinois is the scarlet thread running through Quammen's narrative. Traditionally, the fossil record has been the prime resource for understanding life's progress from simple to complex. Theory proposed that genetic mutations were selectively passed on to future generations resulting in increased adaptive complexity that issued in new species that more efficiently exploited thermodynamic niches. Molecular phylogenetics studies patterns of relatedness found in the molecules of life, particularly ribosomal ribonucleic acid (r-RNA), tracing the exchange of genetic material between species across 3.5 billion years. Simply, r-RNA contains a molecular fossil record that is a far more detailed, less ambiguous, easily compared, and readily accessible reference from live samples with which we can trace relationships and deduce origins.

Infection can be inherited. The human body contains about 37 trillion cells. We carry 100 trillion bacterial cells, plus numberless viruses, fungi, and various mites and worms. We are a community of 10,000 other species. Research reveals that we sometimes trade genes with these other species by horizontal gene transfer (HGT). There is no purely human "I." Thus, Walt Whitman was prescient when he wrote, "I contain multitudes."

Consider an iPad or smartphone. They are just a chunk of metal and plastic, useless by themselves. Like the human body, it is the coded programs and "apps" that make them work. We customize them in all sorts of ways by going to the "apps store" and downloading code to tell us the weather, clean-up our photographs, message others, video the grandchildren … Well, cells can use the same process—they can insert or download packets of genes from other organisms to add new functions through reproduction and/or infection. In reality, we are a composite of "genetic apps" downloaded, modified, and fixed in our software across 3.5 billion years of evolution. The gene to make the enzyme glutaminase is exactly the same in every living thing and goes back 2 billion years. Mitochondrial and ribosomal DNA are non-human DNA and can be traced back through the primates and ultimately to a proteobacterium perhaps 2 billion years ago. Roughly 8% of the human genome is patched in remnants of retroviruses that infected our ancestors.

Code Riding on Starlight

Now, let's circle back to the question posed at this chapter's beginning—what is life? Various suggestions were offered along the way, but nothing definitive, nothing that links the 11-million-year-old mouse with artificial neural networks that may soon be in charge of this Earth. Allow me to offer this little meditation from an evening years ago, standing on a lonely hilltop under the singing fire of the Milky Way:

"What if I am, in some way, only a sophisticated fire?"[8] — Loren Eiseley

Upon retirement I built a lovely home on hilltop acreage we owned in northwest Connecticut. The view was open for miles across undeveloped ridges and valleys. Puffy cumulus clouds drifted horizontal to vision. We watched rain and snowstorms, like slanting gray brooms, sweep ridges far in the distance. But the view was best at night, remote as we were from streetlights and cities, when the great dream-river of the Milky Way spread out across the heavens with stars sparkling like dewdrops coalesced on a cosmic spider web. Pondering the vastness of our stellar galaxy one night, words came unbidden on a wave of emotion, and I uttered, "There is my home." I named the property, "Starfields."

One wonders at the source of such thoughts, where they come from and what secrets of our two-million-year journey under the light of stars are locked away in the hidden interstices of the subconscious mind? For most surely, there are more connections, more genetic memories buried in those shadowy depths than we will ever understand.

The brain contributes 2% to our body weight yet consumes 20% of our food calories, 20% of our oxygen, and is given priority over every other body part. Even in extreme starvation the brain derives energy from shriveling limbs and organs. We have a furnace within our crania producing energy to maintain electrolyte balances and send thought riding on the blue arc of starlight discharging across 100 billion neurons wired together in the most complex parallel network known to exist. Here, in this stew of calcium and sodium, carbohydrates and massive blood-flow, ions and electrical discharges, a vast network holds coded memory of childhood Christmases, the hurts and triumphs of life, the smell of bruised rosemary, multiplication tables, our first kiss, religious beliefs, political opinions, and the taste of good salmon. In his, "Astonishing Hypothesis," Nobel laureate Francis Crick paraphrases Alice by saying, "You are nothing but a pack of neurons." Perhaps. Yet, here is the "I" that we call self, and it is immaterial. Self has no substance. The world's best neurosurgeon could not open a brain and find a self. It appears that self, the "I," is a composite, a story of all our memories, beliefs, sensations, woven together much as the pages of a book are stitched into a singularly unique volume. Our "self" is, in the analysis of cognitive scientist Daniel Dennett, the center of narrative gravity. By an ultimate reductionism, "self" is code riding on starlight.

In memory, I stand once again on that dark hilltop pondering life amid the singing fire of the Milky Way, and hear Norman Maclean: "Eventually, all things merge into one, and a river runs through it." There is no division of source from substance. The trees, the stones, the leaves talking in the night breeze, a family of coyotes howling by the valley beaver pond, silent deer browsing on clover, a man and his thoughts—we are all reconstituted starlight, riding a sphere of reconstituted stardust 8,000 miles in diameter, light and darkness flowing together as one, into a boundless and sacred river of light and life.

8. Truth

"Truth is one, the sages speak of it by many names."

— The Vedas

"Those who can make you believe absurdities can make you commit atrocities."

— Voltaire

"A conscience which has been bought once will be bought twice. The loyalty to humanity which can be subverted by a skillful distribution of administrative sugar plums will be followed by a loyalty to official superiors lasting just so long as we have the bigger sugar plums to distribute."

— Norbert Wiener[1]

"The very concept of truth is fading out of the world."

— George Orwell[2]

Basis for Truth

REJECTED AND DERIDED, Jesus responded to Pilate's questioning, stating that he had come into the world to bear witness to the truth. Mockingly, the Roman governor replied, "What is truth?"

It appears that the spirit of Pilate is alive and well in our world today. The political and economic elite who pull the strings of government have largely disavowed the quest for truth in pursuit of selfish purpose.

Falsehood is multiplying in a new nationalism, by global warming denying industrialists, and even religionists who support leaders and policy that are the antitheses of their founding documents. The assault on science common today among various ideologues is as much an attempt to suppress the truth as it was in the days of Galileo. Religious fundamentalists have become cultist who disavow truth because it undermines their power among the people. Fossil fuel executives suppress truth because it threatens their fortunes. Politicians deny the truth because it would expose their corruption.

But truth is immortal because nature is immortal. We never break the truth, only ourselves on it.

If epistemology (our basis for knowing) is the method by which we validate truth, then we are in a serious crisis. On the one hand, we are the consequence of nature's additive truth. Our cells and systems are derived from eons of sifting—trial and error—on the part of nature to produce logically arranged patterns of information that fit harmoniously within their environmental niche to have the best chance for continuance and reproduction. On the other hand, there is man's truth—a system of logic designed to yield the most efficient way to get what we want with the least effort. Gregory Bateson identified "human conscious linear purpose" as the greatest threat to Earth systems.

Ultimately, some ideologies construct logic chains designed solely to attain conscious linear goals in spite of their long-term consequences to the larger system. We value immediate gain and discount potential future loss.

Conscious short-term purpose, the invention of language, and development of human technologies abstracted us from the world of nature. Nature is the realm of non-conscious processes, truthful feedback mechanisms, and a language (syntax of relationships) that predates all human systems. With the rise of consciousness, we were able to superimpose our systems of language and logic over those of nature to secure selfish immediate purpose. We developed human stories (myths) centered on man to describe our new and abstracted world of language and purpose. These abstract stories are often divorced from and elevate us to a position of dominance over nature. By our dualist myths, we divorced ourselves from a relational reciprocity with nature and took control based on new dominionist narratives.

In Ovid's *Metamorphoses,* Phaethon is the sophomoric son of Clymene and Phoebus who wants to prove himself a capable man. Day in and day out, he pesters his father to allow him to drive the chariot of the sun across Earth. Relenting one day, his father grants permission and Phaethon commands to the horses to be off on their journey. Eventually, however, his weakness and inexperience expose the folly of hubris. He drops the reigns, allowing the horses to carry the sun too close to Earth, burning trees, scorching the plains, and killing life. Eventually, Phaethon is cast out into the sky and blazes like a shooting star until his lifeless body falls to a now blackened earth.

In our hubris we became Phaethon, grasped the reigns, and are now scorching the fertile earth—all based on a contrived narrative set in an abstract world inaccessible to empiricism.

Meanwhile, David Hume (Scottish Enlightenment philosopher and economist) is rolling over in his grave, shouting, "…hearken to no arguments but those which are derived from experience … reject every system of ethics, however subtle or ingenious, which is not founded on fact and observation." Hume saw what Voltaire, Blake, Swift and myriad others have seen—that if people can be enlisted to believe a fabrication, and incorporate that fiction into their world view, then they can be manipulated into the service of those who are the stewards of those fabrications. This fiction is the power of elite ideologues who blow their dog whistles and cause multitudes to fight their wars, demean other races, denigrate the less fortunate, sacrifice for the construction of the temple, defend the "divine right" of the elites to rule over them and violate their own moral code to protect the myth. When the fiction is exposed for what it is, they lose their power. And that is exactly why the powerful will stop at nothing to maintain the sanctity of their fabrications. In every age, people have been slaughtered, burned at the stake, shunned, bankrupt, fired, and crucified because they had the audacity to cast aspersions on the elite's "golden calves."

Honest and intelligent people are often humble. Newton thought of his discoveries as a few pebbles turned over on the shore of a vast ocean. It is this attitude of humility that opens us for new information that can alter previous understandings. As Mary Catherine Bateson has said, "The truth-value of scientific knowledge depends on its openness to correction."[3]

But if our values are derived from some divine order fixed in the heavens, then they cannot be changed unless the God who decreed them changes. And that is a violation of rigid theologies that worship an unchanging God. Thus, if God changes, he is not God because his former pronouncements might be wrong. And if God is brought low, then our story and place in it is called into question. If this God is proven a myth, then what of our self-story? Foundations begin to crumble. This crisis of our time is manifest in culture wars and rising political polarization. Two competing stories are at war—derivative fiction and scientific empiricism—struggling over the basis for truth.

Cybernetics, systems theory, information theory, and ecology have given us new tools to evaluate truth from a far higher vantage point— that of systems of interrelated entities.

These new sciences focus on patterns of relationships.

Science has validity because nature speaks only truth. And that truth is empirical, validated by trial and error between the forces of the universe and laws of thermodynamics. The little hummingbirds that come each day of the summer to our red begonia blossoms seem so delicate, so fragile. But that's my perception and it is limited. In fact, they have been around far longer than man, perhaps 22 million years. And 22 million years can throw an awful lot of stress tests on a species to validate fit and pattern. Man has been around just 10% of that time and we are showing some serious inadequacies.

In the end, truth is manifest by proper fit within patterned relationships for long-term continuity. Truth is combinant. Non-truth is non-combinant. Truth facilitates pattern building; non-truth ruptures pattern because it is illogical. Mathematical formulas containing an illogical element become absurd. They break down upon examination. Neither does evolution lie. Long-term patterns, as manifest in living things, are true because they are logical, and that logic is the trustworthy basis of scientific inquiry. Truth subject to empirical testing, observation, and duplication provides a far more stable platform for a solid worldview than any number of fictions.

Therefore, from nature's perspective I suggest that, **truth is any input that maintains or enhances systemic integrity.** Truth is demonstrable. Truth possesses integrity. Truth embraces light; non-truth shatters upon examination.

Truth and Beauty

The English word, beauty, derives from the Latin, *bellitas*, plus full, to yield—full of beauty or beautiful. I did some digging into a Greek word for beauty, *omorphia*, but could not find an adequate breakdown of meaning. My guess is that omorphia is a compound of *homo* = same, which got shortened, and *morph* = body or form. Thus, I am thinking omorphia means balanced body form or symmetry, because the concepts of balance and symmetry lie at the heart of most understandings of beauty. Since symmetry is such a powerful concept in understanding the universe and how it operates, it seems logical that symmetry often implies a "rightness" quality to things, just as lack of symmetry implies dis-integrity. That which is right is also true, hence Keats' pronouncement that "beauty is truth and truth beauty." Both are righteous because they reflect the basic order of the universe which is a manifestation of the unseen but *Immanent Logic* from which they are derived. Ordered matter is a manifestation of an ordered and immaterial reality. Therefore, Keats saw beauty as a reflection of divine order, or truth. Carl Jung hints at this in his *Memories, Dreams, and Reflections*, 'The earthly manifestations of 'God's world' began with the realm of plants ... They expressed not only beauty but also the thoughts of God's world ... Trees in particular were mysterious and seemed to me direct embodiments of the incomprehensible meaning of life."[4]

The more common Greek word used for beauty is *kalos*. *Kalos* came to the Greeks via the Sanskrit *kalha*, interpreted variously as "sound," "powerful," or "healthy." Thus, throughout ancient Greek literature, *kalos* implies health and an integrated order of both external appearance and internal composition reflective of divinity. The concepts of "the beautiful" and "the good" merge into one with Plato. A soul-hungering for the good and beautiful is the Platonic motivation behind all learning, discipline, and training. The good ultimately binds man, deity, and the world into one. The third century AD philosopher Plotinus attempted to move beyond the idea of symmetry as the foundation of beauty. "The worth of physically beautiful things is in their reflection of the divine order ... The beauty of the things of this world thus reveals the glory, power, and goodness of the spiritual world, so that there is an imperishable bond between all things, between the spiritual and the sensual."[5] Therefore, the experience of ineffable beauty in this world ushers one into the realm of immanent deity. Order is beauty; beauty is truth; truth is God. To contemplate transcendent beauty is to merge with the mind of God. This is life's highest goal—Aristotle's life of contemplation that connects man to deity. Echoing the reinterpreted Platonism of St. Paul, Plotinus prescribes ascetic discipline, an acceptance of death, and wisdom that focuses on the spiritual above the material. This is the same path later adopted and prescribed by Pascal in his *Pensees*.

If the idea of beauty, symmetry, balance, and integration are fundamental to the nature of the world and to the powers of creation, should we not find an appreciation of this in other living things?

Most people are unaware that the theory of evolution did not begin with Charles Darwin. His grandfather, Erasmus Darwin, developed the concept decades before his grandson's book. Additionally, Alfred Russel Wallace sent his own manuscript setting forth a similar theory to Darwin from the jungles of Indonesia in the same year Darwin published *Origin of the Species* (1859). More than twenty years later

Wallace would write, "The only way in which we can account for the observed facts is by supposing that color and ornament are strictly correlated with health, vigor, and general fitness to survive."[6] For Wallace, physical vigor, beauty, aggressive displays, intelligence, and social position are all truthful indicators of mate quality which imply greater survivability in offspring. Therefore, female sexual choice, based on aesthetic assessment as a tool for deriving the true quality of a mate, is a form of natural selection for the evolution of adaptive forms more apt to survive. Beauty is a basis for assessing systemic integrity, or truth.

In the grand picture, as demonstrated by the various ecological sciences, it is the overall web of interconnections that issue in strength, symmetry, vitality, and wholesomeness we associate with beauty. From the environmental point of view, one could say that systemic balance is fundamental to a perceived sense of beauty. What hits us at the "gut" level as beautiful is a perceived rightness of the total system. When the pattern of an organism, a thing, an organization, or an ideology possesses integrity of form and function, we see beauty, wholeness, health, goodness. Harmonious integration is an expression of truth.

We are drawn to natural areas—seacoast, mountain lake, redwood forest, grassy savanna—that grace the Earth as they have for thousands of years, unchanged by the hand of man. One cannot don her raincoat and hike for miles in the Hoh River rainforest (Olympic Peninsula, Washington State) amid hanging moss, ancient spruce, and the smell of salmon bones bleached and rotting along the riverbanks, without sensing an overall rightness derived from the integration of Earth, life, sea, and mountains. Eagles descend to the riverbank to pick at those left-over salmon bones. Steelhead may be migrating up from the Pacific. The air is rich, heavy, and damp with the primal essences of Earth and water and life. Life and the land are as they have been—a partnership—for millennia of time before our kind walked these pathways. And that wholesomeness begins to seep into the hiker. It's not just one facet, but the combined context of a harmoniously integrated and functioning ecosystem that resonates deep within, that tells her at the subconscious level that this part of the world is in proper order, that everything is right in its own place, integrated, and that the best thing one can do is to respect that order and interfere as little as possible.

Once back at her car, our hiker will know that she has brought the entire ecosystem back with her. The wholeness in which she walked has seeped into the interstices of her psyche and body and she can feel it. It is health; it is beauty; it is truth that provides a standard to which she can adjust her life. Presented with nature's integrated wholeness, we become aware of our own imbalances, our own in-sanities. The lesser mind adjusts to the greater mind, and in the process, it is cleansed and restored and reconnected.

Ugliness as Non-Truth

Just as we are attracted to logically integrated patterns in nature we are repelled when exposed to patterns that are broken, polluted, dis-integrated. Unfortunately, billions of people are required to live amid the broken patterns that result from our conscious linear purposes that often disregard the overall health of the system of which we are a part.

We have filled some of the world's most productive wetlands with our junk. Vast, poor areas of cities have streets strewn with garbage and burned-out cars. Industrial communities, never beautiful in their heyday, now sit as rusting scrapyards fallen into dissolution. Cities are crammed with cars and fumes and cruelty. We have released hundreds of millions of gallons of oil and chemical dispersants into the Gulf of Mexico. Species are going extinct faster than at any time in human history. Coral reefs are bleaching and dying. Polar ice caps are melting, threatening to flood multiple major coastal cities

And the root of all this destruction lies squarely in the ideologies that we hold sacred—free market capitalism that has become perverted into a parasitic depletion industry, religious dualism that tells man Earth is not his home but a store of resources to be extracted under his dominance, and a war machine run by a perversion of democratic governance that views other people's resources as our national interests giving us the right to take what we want, when we want it. The dis-integrities of our mental ideological systems are projecting their broken patterns onto the world causing a catastrophic loss of living things and the destruction of vast ecosystems upon which our own survival depends. Our insanities are driving the Earth to insanity. A sick Earth cannot foster a healthy humanity.

When we are exposed long enough to falsehoods, those falsehoods eventually become our reality. For more than 5,000 years, western man has been told that Earth is an allurement to sin and is not our true home. Hitler's minister of propaganda, Joseph Goebbels, once said, "A lie told once remains a lie, but a lie told 1,000 times becomes the truth." We have lived under a system of non-truths for so long we believe it true. But what we are finding is that an illogical system of false ideas can never produce beautiful outcomes.

And yet, once we have accepted a narrative and lived in its territory for a period of time, it becomes home. Few people are willing to leave a known place for the unknown without strong motivation. That's both because we do not like uncertainty and because pioneering new territory is a lot of work that comes with no guarantees. Remember, the unknown is the place where "there be dragons." On top of this, we usually run in tribes that hold the same world view. Tribal views reinforce over time as we find solace in a group that sees the world in mostly the same way. Like Spinoza, we either confront the group with our difference and risk alienation and shunning, or we shut up. Few people can scale such a mountain. So, we end up going along with the group, like the many basically good people in Germany during Hitler's rise and campaign of genocide. Good people become complicit by their silence. It is not easy to speak truth to power. Thus, the lies continue until there comes such a crisis of anomalies that the entire system crashes. False stories are the real murderers behind the guns and bombs.

There is tremendous irony in all of this.

Consider this example: The tribe I spent most of my life with—religious conservatives—had no use (at the leadership level) for atheists, feminists, Planned Parenthood, and homosexuals. We were in a holy war against such people because they could be the source of God's judgement on our society. Those people should not be permitted to have power in the political process because that would inevitably lead to the breakdown of the monogamist family, traditional values, the loss of our exalted roll as world leader, the infection of our children with sinful ways, and the communists would take over as God's agents to punish a society that would not stand up for truth.

For decades, TV evangelists have been walking about in sackcloth and ashes (figuratively), telling us of the doom atheists and the teaching of evolution in our schools would bring on society. Yet, an army of atheists could not hurt the cause of religions more than its own leaders who support politicians and policies that are in complete contradiction to the teaching of their founding documents.

Everything we know, not only from the Bible but from philosophy, other world religions, systems theory, and the ancient Greeks tell us one incontrovertible fact: The very nature of God, however one defines God, is Truth. It is the people who talk righteousness, while supporting lies, who make a mockery of truth. And that derision poisons the well for every honest seeker of truth.

The Aesthetic Response

What I have just said is neither new nor original. John Ruskin, William Blake, Henry Thoreau, Charles Dickens, Martin Luther, Jonathan Swift, Gandhi, Jesus, Hosea, and Frederick Douglass express similar opinions on multiple occasions to their own contemporaries. Those who speak truth to power are generally exercising good logic. Unless we face the pain and suffering that are a result of ideologies built on non-truth, admit all the ugliness which our thoughts had a hand in delivering, we can never move on. No new structure of any lasting worth can be constructed on a rotting foundation. Therefore, the primary job is to put forth better ideas—ideas that can withstand close questioning, ideas that are not full of anomalies, ideas that will stand the rigors of time and change. Should we also take up arms?

A great number of classic children's stories hinge on the theme of good versus evil. The evil villain sets up a response and we want to take up arms for his defeat. Eric Nicol addresses this response in his short parable of "The White Knight": A wealthy knight observes the destruction left in the trail of a black knight going about the countryside sowing all forms of pain and suffering. On the wave of righteous indignation, he dons the armor of a white knight, mounts his steed and goes about searching for black knights. Looking diligently for years, he never finds the black knight, only those who are shades of gray. Ultimately, after years on the road, he looks up to see another white knight. But this one is bearing down on him with leveled lance. By a defensive response he ducks the lance and thrusts his sword into the white knight, mortally wounding him. Stepping from his horse, he sees under the raised visor of the white knight a young, golden-haired boy set out to fight evil. With his dying breath, the boy asks, "Is evil then triumphant?" Beside the boy's white armor, he could see that his, in the darkness of the forest, looked black. He buries the boy, then returns home to clean himself, dress in multicolored clothes of cheerful taste, and never again looked for more than he could see.

The obvious moral is that one cannot fight evil with the weapons of evil and arrive in a good place. On a societal level, however, we still do not grasp this message. We engage in war to end war. We attempt to enrich ourselves by impoverishing the Earth. We take to the political process with all of its surreptitious workings, character attacks, and accommodation of the powerful, thinking that we can topple the system using the tools of that same system. In the end, we become that which commands our greatest attention. Engaging evil on evil's terms only produces evil. How it happens I do not know, but we tend, by an inexorable law of the universe, to become the thing we are most focused on. Darkness can only be set to flight by the rising sun.

If we want to cultivate truth, we must work in the fields of beauty. Truth is a friend of light. Evil flourishes in darkness. If the aesthetic response to truth is a sense of beauty, then those who desire to cultivate wholesomeness cannot exercise the tools of war, because they contradict the mission of truth. The Christian Right, seeking to defeat a more liberal world view, has engaged the tools of the political process and is currently supporting several "black" knights in an effort to make society "white" again. This is a contradiction of values that exposes the effort as just another power grab. In the process they are sullying themselves and have become, in the eyes of many observers, the evil they are seeking to vanquish. Goodness overcomes evil; beauty satisfies; light vanquishes darkness.

Perhaps this is why Gregory Bateson refused to get involved in politics. He realized that the ugliness derived from conscious linear human purpose cannot be defeated by legislation alone. He was convinced, especially as he aged, that only a religious-type response was adequate to save us from global ecological disaster. Although an atheist, he viewed the interconnected whole of Earth systems as "the sacred." It is the aesthetic response to beauty, which is innate in all people, that has the most power to turn us away from the ideologies of destruction to a new way of living—the way of beauty, harmony, and belonging.

On the whole, people tend to figure things out over time, especially when exposed to personal pain and loss. As global warming and accompanying political instabilities rise, pain will also rise. Whether this will lead to war and genocide or to coordinated efforts to eliminate fossil fuel production and benignly limit human population remains to be seen. Working for, communicating, and practicing the good are always the best paths to permanent change.

Psychologists are learning that most of the things that make for a good society are pre-programmed algorithms already built into the human brain across eons of evolution, and they are not unique to humans. Chimps and bonobos are also programmed for cooperation and reciprocal altruism. The upshot is that we are all built and pre-programmed to survive on the basis of mutual benefit derived from cooperative behavior. In the end, I hope this desire for survival and mutual altruism will overcome the destructive tendencies that are making our world poorer for survival and reproduction ...

The desire for life supporting habitat is pre-programmed into every brain. Over a two-year period, I looked at more than one hundred properties, seeking a place to build our zero-net-energy retirement home. I walked through forests, climbed mountain tops, looked and looked and looked. In my mind, however, I had this image of a forested point of land viewing a large, open wetland system. Driving by a similar place one day with a realtor I pointed and said to the realtor, "That's what I'm looking for." It was not for sale. More than a year later, however, another realtor sent a new-to-market listing to my email. With the listing were pictures of this forested land surrounded by a grassy meadow and 100-acre pond against a backdrop of mountains. At the time we were 1500 miles away, in Florida. But on seeing those pictures I said, "That's the one." And here we are eight years later living the vision.

The Adapted Mind has a chapter titled "Evolved Responses to Landscapes." The thesis proposes that we respond emotionally to landscapes that are positive for survival and reproductive success. Evolutionary biologists have long believed that modern man and his predecessors originated in the savannas of Africa. With their long, open views, abundant wildlife, and variety of vegetative foods, savannas were an optimal place for security, survival, and reproduction. Tests in which people were shown various habitats—savannas, deserts, mountains, and seashores—found that eight-year-old children preferred the savanna ecosystem over all others. It seems that younger children would choose more from what is built into the brain by evolution as opposed to later exposure. This is the *Savannah Hypothesis*.

Additionally, we prefer spaces that protect us overhead, offer long vistas, and provide protection at the back and sides. Does this evolutionary programming lie behind the appeal of Frank Lloyd Wright's Fallingwater? I have toured that house, originally built for the Kauffman family, several times and the attractions are subconsciously powerful.

Shortly after we finished our home a friend came to visit. Looking over the vast marsh/pond complex out front, with the hundreds of acres of sedges, grasses, pitcher plants and cranberries, she remarked, "Wow, you built right on the Serengeti."

Here is the question: Does this land speak only to me? Or does it speak to the thousands of human generations and their experience that yet live somewhere deep in the labyrinthine tunnels of my neurons? Are we not one in a continuum? And if so, is not all mankind now on Earth one? Perhaps this is the meeting place for a common truth that will rebuild a world where we can again be secure in our persons and our perpetuity. Beauty leads to truth, and truth leads to healthy integration which facilitates long-term continuance.

Wabi Sabi

For decades, I have been a lover of Japanese gardens. At our former home in Connecticut (Starfields), I used our 22-ton excavator to make a Zen garden of about one-quarter acre in size, covered in light gray raked gravel, and with seven large and strategically placed stones. It was the first thing we saw each day as we sat with morning coffee. And it was hauntingly beautiful. When the property went up for sale, the first person to come bought the house largely because of that Zen garden.

Upon completion of this cabin, I wanted that same understated Japanese aesthetic to embed structure with Earth. Sorting through my stone pile, I found just the right moss-covered stones to create a small Zen garden against the north wall. Additionally, another one-ton stone turned into a carved wash basin. Three others, plus four granite legs, became a three-ton Shinto lantern set opposite the wash basin that one must pass before climbing the stone steps to the cabin. They bind me to an Earth that is ever passing and ever being renewed.

The Japanese concept of Wabi Sabi reminds us that life is imperfect, impermanent, and incomplete. Golden ginkgo leaves fluttering earthward on a late October day, evanescent cherry blossoms yielding their mortal beauty to the mud of spring rains, mosses and lichens slowly chewing away at an ancient stone—each creates a serene melancholy and spiritual longing for the fleeting perfection implied in a transient beauty. They become portals for meditation on truth—the perfect, the immortal, the unchanging. They remind us that life is ever renewed within the golden circle of time and the universe—that the flaming leaves of October must scurry before the winds of winter so they can return again, on some far-off night in May when the once frozen marsh metamorphoses into the euphoria of spring peepers' "Ode to Joy."

9. Broken Patterns

"Carbon-fueled capitalism is a zombie system, voracious but sterile … It is unsustainable, both in itself and as a response to catastrophic climate change."
— Roy Scranton[1]

"…you can draw a very credible climate connection to this disaster we call ISIS right now."
— Rear Admiral David Titley, retired[2]

"Right now, we are in the midst of the Sixth Extinction, this time caused solely by humanity's transformation of the ecological landscape."
— American Museum of Natural History, plaque

"This demand for secrecy is scarcely more than the wish of a sick civilization not to learn of the progress of its own disease."
— Norbert Wiener[3]

Runaway

IT WAS A HOT SUMMER NIGHT. I was 12 years old and had stayed up late with my parents to watch the 11 o'clock news. That's when we heard the blaring horn of a runaway truck. Our home was at the mouth of a hollow rooted to the backbone of Chestnut Ridge in southwestern Pennsylvania. Route 40 ascends the 3.5 mile incline to the top of the ridge, where there is a huge sign with flashing lights telling truckers to descend in low gear only. Occasionally, an unfamiliar trucker on a tight schedule would disregard these instructions, burn-out his brakes, and crash.

Within seconds, we heard the awful crash as the speeding truck failed to negotiate the turn below our home and careened into the trees. I outran my parents to the crash, just as our neighbor, Mr. Dillow, arrived. The truck was overturned with a crushed cab. The driver, however, was quite alive and screaming for help. In spite of a diesel fire, Mr. Dillow ran to the cab and grabbed the man's arm. But the cab was so deformed he could not extricate him. Meanwhile, the fire increased to a blaze that singed my neighbor's hair and forced his retreat. Then it happened. The fire reached the cargo—nitrate fertilizer. The fertilizer erupted in a blaze that reached the treetops, roaring. I will never forget the driver's dying screams …

The fact of anthropogenic global warming would not be so threatening if it were a linear process which we could reverse with existing renewable energies. Nor would skiers be in grave danger if high mountain snows broke off a shovelful at a time. But avalanches come suddenly and grow exponentially, after the weight of snow mass reaches an irreversible tipping point. It is the same with avalanches as with fertilizer fires as with global warming. Ice shelves are collapsing in Antarctica, tundra permafrost is melting at an alarming rate, preparing for the release of vast stores of methane (the Methane Bomb) that has 25 times the warming effect as CO_2. Meanwhile, the Arctic Ocean is warming rapidly and could cause the release of deep-sea methane hydrates. Greenland is melting at increasing rates due to warmer weather and decreased albedo. Himalayan glaciers are in retreat portending vast food shortages in southeast Asia. Forests are being clear-cut worldwide. Climate change is impacting agriculture in Africa and the Middle East. Earth is now in its sixth Great Extinction.

We are rapidly approaching a tipping point where positive feedback loops reach an irreversible runaway state. When that occurs, scientists expect temperatures to rise by more than 5-9°F by the end of this century. Oceans may inundate coastal cities, leaving up to a third of Florida underwater. Food production may drop precipitously. Tropical and subtropical areas may endure months of temperatures over 100° with heat indexes in the 130°s. Millions of refugees, famines, and financial breakdown will leave civilization in chaos. At this point methane will release faster and in larger quantities. This will cause even more ice to melt as temperature rise accelerates. Every living creature will be fleeing poleward, fighting for food and water and safety. With the thin veneer of civilization peeled and bleeding, we will have entered the world of Mad Max.

A recent report in *USA Today* stated that 99.999% of scientists believe that global warming is real and that it is caused by human generated greenhouse gases—carbon dioxide and methane. Climate scientists have measured, recorded, statistically analyzed, and mapped pollutants and weather events to the point where there is no longer any doubt that we are wagering ourselves into a dystopian future. Continuing business as usual, average temperatures may increase by .16° C/year in the Arctic. *Yale Climate Connections* predict that such an increase would result in a runaway release of methane now locked in tundra permafrost. If this rise continues for another 20 years, the Methane Bomb will have detonated and there will be nothing we can do to stop it. Alaskan temperatures this March (2019) were 20°F above average.

Although the specific facts are different, the ecological, religious, cultural, and economic stresses now at play are following the same form Thomas Kuhn outlined over 50 years ago in his *Structure of Scientific Revolutions*. The old systems, assumptions, and points of view have lost validity and stand ready to be replaced by fresh insights, a broadened sense of justice, and demonstrable truth.

For 3.5 billion years, evolution has been at work building ever higher and more complex patterns as life systems with a multitude of amazing emergent qualities. We are now standing on the threshold of a dystopian period that threatens to undo in centuries what it took evolution eons to accomplish. We are in *The Age of Depatterning*. And we are part of the pattern.

Lies, Malware, and Logic Bombs

Beliefs founded on an illogical basis tend to foster illogical practices that result in broken patterns. When **Dualism** abstracted God from Earth and installed him in an unseen heaven, it left a void. Man stepped into that void to take control as the vicar of God over Earth. **Dominionism** was born. By assumed divine right, man took dominance over Earth and women, who reflect Earth by their fertility. Conscious human linear purpose supplanted nature's non-conscious systems' processes developed by selection and adaptation. And fear of **Death** kept us in bondage to these destructive falsehoods.

Runaway chaos will permit no reversal of self-reinforcing systemic feedback. We may develop a vaccine for Covid-19, but there will be no vaccine against climate change. Vast patterns that were once logically

sustainable will begin to dis-integrate. Systemic illogic is "in-sanity"—
mind that has lost control.

We are currently experiencing multiple crises in racial justice
and integration, economic stress from financial inequality, a struggle
between fossil fuels and renewable energy, populist politics and the
visions of Abraham Lincoln and Thomas Jefferson, fascism and
democracy, and science and religion. These are not stand-alone issues,
but multiple expressions of one thing: The struggle between the old
3-D paradigm and a new model struggling to carry life forward to the
next higher level of evolution based on sustainable integration.

In 1950, Norbert Wiener published *Cybernetics and Society*,
outlining his attempt to understand the logical control of systems
as they struggle to maintain organized and complex patterns while
resisting the disorganizing tugs of heat death and increased entropy.
He described the agent of disorder in terms of the old metaphor of
the devil. But what kind of devil is this? "Is this devil Manichaean
or Augustinian?"[4] The Manichaean devil is a sinister force, willing
to lie, deceive, and sow confusion in order to win his point. This is
the same devil described by Jesus of Nazareth two millennia earlier
as "a liar and the father of lies." On the other hand, the devil could
be Augustinian in nature—not a conscious deceiver but a projection
of our own weaknesses and ignorance. This devil can be overcome,
not necessarily by craftiness as much as by persistent effort and hard
work to uncover the hidden secrets of life and nature. By this effort
we have unlocked the inner workings of the atom and the structure
and function of DNA. We were not in a contest with some evil force,
but with our own ignorance of the intricate nature of the structure
and functions built into complex patterns across the history of the

universe. The struggle against the Augustinian devil is the scientific process of theorizing, searching, testing, demonstrating, and validating results.

On the other hand, the Manichaean devil is "refined malice."[5] He is out to win at all costs, no holds barred. He will obfuscate, trick, deny, deceive, lie, sow lies, threaten, and attempt to destroy in an effort to win. In many ways, the Manichaean devil is the embodiment of Gregory Bateson's "conscious linear purpose" that is focused only on its own goals, no matter the collateral costs to surrounding systems. The Manichaean playbook is currently in use by the vested interests of the fossil fuel and industrial/military complex to accomplish their goals of extractive profit and dominance at any cost. The result is broken patterns. Consider our mismanagement of topsoil—that thin layer of earth that feeds us every day.

Broken Patterns

Soil is the placenta through which we draw the nutritive transforms of Earth's elements and sun's energy. Iowa is the gold standard for farming soils in the United States, with raw farmland now going for over $10,000/acre. When serious farming began, around 1860, Iowan topsoil averaged 13" in thickness. Today that same topsoil is 6.5" deep. Soil scientists agree that farms across North America lose an average of five tons of topsoil/year/acre. A sudden downpour in Iowa, according to a 2011 piece in the *Huffington Post*, can strip about 64 ton/acre. A 2006 article in the Cornell Chronicle estimates that global soil erosion is occurring at 10-40 times the rate of replenishment— equal to an area the size of Indiana lost each year. It takes 500 years to build 1" of topsoil. In an interview with the World Economic Forum, John Crawford, a scientist at the University of Sydney, alleged that we have just 60 years of topsoil left, globally; 40% of that soil is degraded, and that over the next 20-50 years there will be 30% less food while demand grows by 50% due to rising standards of living and population increase. This suggests vast social and political instability.

Meanwhile, a report in *Lancet* calculates that there will be 529,000 deaths/year by 2050, mostly in southeast Asia, due to food shortages resulting from global warming. By 2100, the Central Valley of California will have 50-100 days/year of temperatures over 100 degrees, while Dallas will have 133 days/year over 95 degrees (Climate Central figures). Current wars and civil unrest, along with the attendant refugee crisis in the Middle East and Africa, are a direct result of weather changes, population increase, and decreasing amounts of food. China, which is losing 1,300 square miles/year to desert, is currently acquiring millions of acres of agricultural lands in Africa and South America. Political and religious rhetoric aside, all wars, over all of time, are and have been about access to and control of resources and markets.

A 2013 article in *The Atlantic* described how chemical farming is destroying the soil microbiota that live symbiotically with plant roots and provide water and nutrition necessary for growth. There are more bacteria, fungi, and protozoans in a spoonful of soil than people on Earth. As a result, foods today contain more toxins and less nutrition, soils are more compacted, have less tilth, less natural fertility, hold less water, and are becoming more saline. Meanwhile, between 1995-2014, US taxpayers subsidized corn farmers with $94 billion (*The Economist*), with much of the corn going to ethanol to drive our cars, while fertilizers and pesticides ran into surface and subsurface waters.

Correctly understood, therefore, large-scale industrial agriculture that depends on synthetic fertilizers, genetically modified crops, Roundup herbicide (along with 2,4-D used in agent orange), a host of petrochemicals, and giant machinery is more accurately a form of agri-mining, because it is an extractive industry that, over time, depletes the resource being extracted. Soils, farm families, communities, taxpayers, and nutritive values of foods—all are being depleted under the current corporate/government-controlled system. No civilization, in all of history, has independently survived the depletion of its soils.

This is the essence of *Dirt: The Erosion of Civilizations* (2007), David Montgomery's sad chronicle of civilization's continuing rape of Earth. In his *Collapse* (2005), Jared Diamond expresses a similar viewpoint. And the breakdown affects more than man.

The United Nations released (May 2019) an ominous report detailing the state of life on Earth, stating in unequivocal terms that 1,000,000 species of plants and animals are on the verge of extinction. This is *The Sixth Extinction* (2014) described by Elizabeth Kolbert. Living things are going extinct at a faster rate than at any other period of human history, and many of those extinctions are directly related to human activity. Chairman of the panel, Robert Watson, worries that the foundations of our economies, livelihood, food security, health, and quality of life worldwide are all at risk.

National Geographic magazine recently (Feb. 14, 2019) highlighted a potential "insect apocalypse" detailed in the *Journal of Biological Conservation,* stating that 40% of all insect species are in decline and could go extinct within decades. Loss of insects would cause many ecosystems to collapse. Most species of freshwater fish depend on insects for nourishment. On a recent trip through California's Central

Valley we saw thousands upon thousands of honeybee hives set in orchards to pollinate nut and fruit crops. These same honeybees upon which billions of dollars' worth of agricultural crops depend are also under attack from colony collapse disorders. The world's oceans are in no better shape.

Tropical forest loss (80,000 acres/day; 30,000,000 acres/year) to make room for more grazing, farming, housing, fuel, and monoculture palm oil plantations is increasing rapidly. Vast tracts of the Amazon, often called "the lungs of Earth," are currently being clear cut. Indonesia is also losing hundreds of square miles of forests to lumbering and agriculture. As human populations continue to grow exponentially in these areas, resource extraction must also accelerate to provide for more people who are increasingly desirous of a more Americanized and affluent lifestyle.

Forests are also declining rapidly in the American west, but for different reasons. A warming climate and protracted drought have stressed vast sections of pine and aspen forests to the point of collapse. The once golden groves of aspen covering hundreds of square miles in the Rockies are turning sick and dying. Ponderosa, pinyon, and lodgepole pine are not getting enough moisture in much of their range. Various species of bark beetles smell this stress and attack the dying trees by boring holes and laying eggs. When the larvae hatch, they begin feeding on the inner bark of the tree, effectively girdling it. Traveling around Santa Fe, New Mexico a decade ago, I saw entire mountain sides of pinyon pine that had turned orange in the last stages of death. Since pinyon nuts are a main pillar of the food chain, multiple species of birds and mammals are also in decline.

I am fearful of losing half of the forest that surrounds this cabin.

As climate warms, woolly adelgids (similar to aphids) are moving north. Upon arrival, they will begin to feed on the needles of our hemlock trees, weakening them over a period of a few years, until they ultimately die. Large sections of our mountains that are green today will, within the next decade or so, be swaths of brown sickness scarring the land. Trees that have been an integral part of New England, like the American chestnut of the prior century, will be gone—all because man introduced illogic, often unintentionally, that broke the patterns of once stable systems.

Everywhere, from ocean fisheries to declining polar bears, from grounding whales to diminished river systems, from melting glaciers to fragmenting polar ice—multiple threads of the same fabric are now unraveling because of conscious human purpose that shows little regard for the natural order of things.

Enslavement of Earth and Man

Those who value an alternative point of view would do well to read aforementioned Yale professor James Scott's *Against the Grain* (2017). Scott's overall thesis is that man domesticated grain, and grain (like fire) domesticated man. Although agriculture is generally viewed as a huge progressive step, Scott questions if it did not do great harm. Wheat was domesticated about 10,000 years ago in what is present day Syria. Yet, men continued to live in small bands of hunter-gatherers, dabbling lightly in domesticated crops, until the first city-states emerged in Mesopotamia. With its large marshlands, Mesopotamia was a cornucopia of wildlife and edible plants year-round. Abundant water made it easy to add patches of cultivated grains for supplemental food. By 3200 BC, Uruk was the largest city on Earth with between 25-50,000 people, mostly slaves.

In most ways, life in city states was a regression for the laboring class. Archeological remains indicate that domesticated people were smaller, had thinner bones and teeth, were "deskilled," and show nutritional stress because the cereal grains which formed the bulk of their diet lack essential fatty acids and inhibit iron uptake. On the other hand, hunter-gatherers (who were a bane to the state because they couldn't be taxed) were several inches taller, stronger, healthier, and more skilled. Excavations at Abu Hureyra in Syria yielded 118 different plant species consumed by hunter-gatherers. Thus, these people referred to as barbarians (from Greeks who mocked their speech as "ba-ba") were freer, worked less, had a wider ranging diet that varied with the seasons and animal migrations, had a broad range of skills necessary for survival, had tighter social alliances, more rituals, were healthier (cities were hotbeds for diseases), and could generally provide all their needs with a mere 15 hours of labor/week (see James Suzman's 20-year study of the Bushmen of S. Africa).

Acquisition and control of people is at the center of statecraft, both by enslavement and control of women's fertility. States cannot exist without the endless labor of a working class. Agriculture is intensive, backbreaking work. All of the early infrastructure was constructed by coerced labor (like the columned estates of our antebellum south). City walls served more to keep slaves in than barbarians out. Slaves were the main commodity of trade. Life was drudgery, impoverishment, dehumanizing cruelty, broken families, and epidemic diseases from crowded living and poor nutrition. The Gallic Wars yielded another 1,000,000 slaves for Rome, whose entire population was from 1/4 to

1/3 slaves. Scott estimates that as late as 1800 AD, 3/4 of the world's population existed in some form of forced labor.

All of this calls into question the meaning of the word "civilized." If "civilized" is defined by ethical treatment of our fellows and Earth, who, then, is most civilized? Reflecting on the European hegemony of his time in *Democracy in America*, Alexis de Tocqueville concluded, "... the European is to the other races what man himself is to the lower animals; he makes them subservient to his use, and when he cannot subdue, he destroys." These are the logical results of dominionism.

Patterns of enslavement for extraction and depletion of both man and Earth continue to this day. Although we have largely replaced forced slavery with wage slavery, the plight of Earth's laborers is little different from former ages, and in some cases worse. But the end results are still the same—broken families, broken bodies, dying ecosystems and an increasingly depleted Earth, with ever more of the profits going to a small segment of the ruling elite who free-ride on Earth and their fellows. Increased wealth buys more political and religious power to ensure that more wealth will follow. Meanwhile, Earth and human society become weaker and less resilient by the day. Systems are failing. The social fabric is fraying as nationalist politics sets us against them, friends against enemies, citizens versus foreigners. Species are going extinct. Earth is breaking out in a fever. The seas are dying. Forests are disappearing. A hand is beginning to write a portent on the wall of humanity's consciousness.

Broken patterns are proof of syntactical irrationality. Enzymes to digest protein are a mismatch for the breakdown of carbohydrates. White blood cells do not bond with oxygen. The sperm of a zebra cannot successfully fertilize the egg of a dog. A 20-thread nut will not fit on a 12-thread bolt. These are systemic irrationalities because the syntax is improper—they do not fit one with the other, therefore there is no unified connection to produce meaning. Integrity of fit, like the proper meshing of gears in a machine, provides a sense of goodness—a rightness of relationship—that both makes sense and makes things work.

When Paradigms Break Down

No false, twisted, or bad idea, in all the history of Earth, has ever been destroyed by use of physical force. We have fought wars against fascism, communism, national socialism, paganism, and a multitude of religions—all to no avail. Why? Because immaterial ideas cannot be eradicated by physical force. They are coins of different realms. No number of bombs can eradicate a mental concept. Bombs kill carriers, not concepts. Moreover, killing people for their ideas and beliefs can only be effective if we kill every last person who was ever exposed to the logic of that idea. Killing only a portion of the ideologues serves only to fertilize the entrenchment of the idea. That's why ideological wars are and always will be counterproductive. Bad ideas never die. But they can be replaced.

Wordsworth, Thoreau, and the painters of the Hudson River School, as well as Dickens, Blake, and Ruskin were eminently aware of the crisis set off by the Industrial Revolution and devaluation of man. Many hearkened back to a simpler time of rustic retreat within the breast of romantic nature. But one can never go forward by going backward, according to Tom Wolfe. It's the "Lot's-Wife-Syndrome," in which we petrify rather than progress.

In the early decades of the 20th century, World War I blew humanist's dreams to smithereens. There would be no egalitarian society in a dog-eat-dog world marching to the tempo of the newly ascendant and mechanized industry where time is money and people are resources. Huge sums of money were there just waiting to be made by newly invented industrial technologies. It was the dawning of the age of the automobile and plastics and the lifestyles of Hollywood and the Hamptons. If we had to go to war to secure the petrol to run this new society, then so be it. The powerful were on the hunt for more resources and nothing would stop them, not even war and genocide.

Max Weber, perhaps the greatest sociologist of the 20th century, saw this and lamented that the world had become "disenchanted." Human dreams, once full of idyllic promises of a new age free of drudgery and want, of equality and brotherhood, were mangled in the trenches of war and industrial malaise that slowly evicted the nymphs and wood-spirits who retreated ever deeper into the mountains and howling loneliness of desert places. Romantic art, with its luminescent mountains and thundering cataracts, would be replaced by the divided psyches and horrified screams of Picasso and Edvard Munch. Nietzsche, always ahead of his time, saw this coming and declared the God of Western Culture dead. The paradigm that imposed a separation of man from Earth, dominion of man over creation, the acceptance of misery in this life for salvation from death in the sky by and by—that paradigm, and the imagined God who decreed it, was a fantasy. Could a new paradigm be born, and with it a redivinized Earth where man would be home once again?

Fragmented Metaphors and Bifurcation

We are currently in the greatest period of de-patterning since an asteroid or meteor struck the Yucatan Peninsula, forming the Chicxulub crater 66 million years ago. In a September 2017 article printed in *Science Daily*, MIT professor of geophysics Daniel Rothman calculated that a business-as-usual release of carbon dioxide into the atmosphere would put us at a potential tipping point around 2100. This could result in a mass extinction event similar to the last five great extinctions of life on Earth.

What initially begins as depatterning in the natural world may inevitably rise through the various species of life until it reaches man. Every day on the news there are reports of rogue weather events of an increasingly violent nature. This week alone (May 2019) at least 100 destructive tornadoes have ripped through the South and Midwest, killing people, overturning semi-trucks, flooding farm fields and homes in river plains, and shattering buildings like toothpicks.

As climate warms, once tropical diseases such as malaria and dengue are working northward with advancing anopheles mosquito populations. Here in New England, fishermen who normally made their catches off the coast of Gloucester, Massachusetts are fishing farther to the North as fish stocks move poleward in a warming Atlantic Ocean. As the seas warm, they are also becoming more acidic because carbon dioxide combines chemically with sea water to form a mild solution of carbonic acid. Although the decrease in pH is small, it is enough to weaken and dissolve the calcareous shells of crustaceans that are the basis of the food pyramid for species from salmon to whales. Hundreds of millions of people depend on protein from the sea. Stocks, beginning with the loss of the codfish industry in the Grand Banks, continue to deteriorate. Bangladesh recently imposed a 65-day moratorium on fishing.

For thousands of years, mangrove forests mitigated the effects of hurricanes along the Gulf Coast of North America. But the oil and gas industry, with the blessing of the Army Corp of Engineers, saw fit to eradicate large portions of mangrove to facilitate offshore exploration and shipping. As a result, when Hurricane Katrina struck off the coast of New Orleans, their front line of defense was no longer in place, allowing an unprecedented storm surge to inundate the city. Houston suffered severe flooding two years ago when a storm system came in off the Gulf and stalled over the city, dropping upwards of 40" of rain in three days. Oklahoma, which depends on the extraction of fossil fuels for its livelihood, has been one of the worst-hit states for tornadoes. That same Army Corp of Engineers also green-lighted the leveling of over 1,500 mountain tops in West Virginia for the extraction of coal. Stream valleys were filled with the overburden, wiping out entire ecosystems. The quality of life for people whose families live among those mountains has been decreased as the old culture is buried under fumes from explosions, dust from trucks passing non-stop, and rubble from the bowels of Earth. Meanwhile, the elite just cannot seem to understand why an opioid epidemic is running rampant ...

In 1900, the human population of Earth was 1.6 billion. Today that number stands at 7.4 billion, headed upward to 11.2 billion by 2100 according to the United Nations. Politicians, however, decry shrinking populations in places like the US, northern Europe, and Japan, saying we need more people to make more stuff so we can make more money. It's like a mantra to the wealthy that they always need more 'human resources' to keep the costs of labor down. And yet, there is no way the Earth can continue to support this many people at an acceptable level. War, genocide, starvation, or infectious diseases will ultimately cull excess numbers until population levels are in balance with Earth's carrying capacity. We can engineer a soft landing through public policy for low birth rates, or we can choose a hard landing, if not a crash, by ignoring statistical evidence and historic facts. Paul and Ann Ehrlich (*The Population Bomb*, 1968) were not wrong, just premature.

Demented

According to cybernetics, all systems are mental in nature—they are minds capable of understanding and processing information for their continuance, communicating information by feedback mechanisms, to operate in a dynamic range near homeostasis. To be successful, these same systems must order information in logical patterns. Systemic illogic functions as faulty code sending erroneous feedback into the system causing maladaptation and potential breakdown. The result is a broken mind incapable of processing information for continuance. Its parts no longer speak the same language. The system is demented—it has gone insane.

In the ancient world, sickness was associated with uncleanness. Hence, when a leper walked the streets of Jerusalem 2,000 years ago, he was required to call out, "Unclean, Unclean," as a warning to others lest they become infected. The association of "unclean" with "unsanitary" is reasonably clear. Thus, a system that is corrupted is "unclean," is not "sane" (sanitary), is not "whole or healthy." When all the parts of a system are communicating, it exists in balance. Faulty communication becomes evident by imbalance. Imbalance is illness. Illness within systemic communication is insanity—internal mental operations can no longer synchronize. One part of the mind is telling another part something that is either distorted or not true. The system becomes confused or is left in the dark and can no longer maintain control. If the governor of an engine stops giving honest feedback, the engine may accelerate to the point of destruction or shut down. Mind has been compromised.

In the case of living things, when the message breaks down the mind breaks down. Message is command and control. Broken or distorted message is loss of control.

In response to the question posed by *Edge* in 2017, "What scientific term or concept ought to be more widely known?", Brian Knutson, Stanford professor of Psychology and Neuroscience, responded with the idea of "Future Self-Continuity." The concept of Future Self-Continuity (FSC) asks individuals to project their lives five years hence and give an evaluation. Any projection into the future is merely a probability assessment. We base our outlook on current information analyzed on the basis of past experience built into our minds as logical patterns. If new information is coherent with past experiential patterns, expectations are for life to continue as normal. On the other hand, if the world has become an unstable place with increasing information which does not logically integrate with established mental/ecosystemic patterns, then future probabilities become uncertain, if not questionable, creating doubt and insecurity. Research indicates that individuals with lower future self-continuity display more addictions and suicides. Knutson notes that, "...cultures that value respect for elders tend to save more, and nations with longer histories tend to have cleaner environments."[6]

Informational patterns have a way of showing up as mental and bodily effects. When the evolutionary patterns of the mind no longer synchronize with the patterns of our environment, systemic communication and reciprocity break down, as manifested by ugliness, sickness, and social collapse. This systemic dementia is a hallmark of our time. Broken patterns and broken minds are mirror images.

Visions of Madmen

"Talent hits a target no one else can hit;
genius hits a target no one else can see."
— Schopenhauer, 1819

William Blake (1757-1827) was a troubled spirit, often overcome by paranoia and schizophrenia. Blake was torn between a deeply religious vision and the blatant hypocrisy of the church. He sought release from the "mind forged manacles" of false ideologies and self-serving beliefs that bind us, that we might "Live in Eternity's sunrise." He railed against churches that condoned child labor as a moral obligation, but refused entrance to the tiny, filthy chimney sweeps. He excoriated the rich who lived in opulence built on the blood of soldiers and indigenous people. Blake was equally at home leveling his charge against the scientific view of man and the world, "May God us keep, from single vision and Newton's sleep." He could never quite square his heavenly vision with a coal-blackened, Dickensian industrialism.

John Ruskin (1819-1900) was the foremost critic of art and architecture in England, wrote volumes on art history, political and economic theory, and taught at Oxford. The man was certifiably insane, having experienced seven breakdowns, and lived the last ten years of his life in silent madness. His wife of seven years sought divorce from an unconsummated marriage. Ruskin believed great art was the product of a just social order based on the command to love one another. Truth induces the aesthetic response of beauty; ugliness is a recoil from falsehood. He believed that architecture is an expression of life and character, and that a person's taste is an index of his morality—"Tell me what you like, and I'll tell you what you are." He refused to design a new Exchange, believing that trading is thievery based on asymmetric strength and information. Said Gandhi, "(Ruskin)...transformed my life." Proust wrote, "...is he not to some degree the Truth?" For Tolstoy he was "one of those rare men who think with their hearts..."

I submit that the greater fault was not in the minds of these and others who reacted to the devastations of their times, both social and environmental, but in the minds of the masses who had built empires on corrupt doctrines and beliefs. The insanity of willful blindness was built into a system based on superstition and imagination.

It has become abundantly clear by now that the problems our world faces today—racism, gross financial inequality, environmental pollution, populist and nationalist politics—are each facets of one systemic problem. And that problem is false information that issued in a cultural religion of dualism, dominionism, and death that has paved the way for the suicidal insanity that is now destroying life forms and systems of Earth based on a false narrative ...

Today's work is finished. I will go shortly and inspect the pink lady slippers that grow along the marsh-side path I take each morning to this cabin. They first showed green just a week ago, pushing up through dried oak leaves. Flower buds were nodding from extending stalks when I returned from lunch. In another week the pink-veined slippers will unfold their sensuous labia, blushing lady-like in the morning sun. How delicate, how sensitive they are, demanding only the best sites for filtered light, mutualistic fungi they share with wild blueberries, moisture, and soil acidity. Their sensitivity is an assurance. Each time I pass and kneel in admiration they give me subconscious comfort. The wholeness of their delicate pattern is a message, an affirmation that life's logical patterns will somehow continue, in spite of the current depatterning, to a new day that dawns ever closer to "Eternity's Sunrise."

10. Feeling

"And I have felt, A presence that disturbs me with the joy, Of elevated
thoughts; a sense sublime, Of something far more deeply interfused,
Whose dwelling is the light of setting suns, And the round ocean and
the living air, And the blue sky, and in the mind of man."

— Wordsworth

"Probability theory is nothing but common sense
reduced to calculation."

— Laplace

"There is no wealth but life. Life, including all its powers of love,
of joy, and of admiration."

— John Ruskin

Salmon Sunday

IT'S ONE OF THOSE BEAUTIFUL, crisp, clear blue-sky New
Hampshire days in mid-October with golden maple leaves wafting
earthward on the slightest breeze. A small field is full of parked cars.
Friends and neighbors, clad in boots and fleeces, are milling along in
the general direction of a silver veil of water pouring over the high
spillway of Mill Pond. Waters dance and boil into foam at the base of
the falls, then move along in a current toward Lake Winnipesaukee,
where they will mix with other waters in a 72 square mile basin on

their seaward journey.

But while people are moving toward the spillway, and waters are flowing toward the sea, another force is working its way upstream. Landlocked salmon have lived and fattened all year in the depths of our lake. Now, responding to an innate and primal urge, they are fighting their way upstream to deposit the beginnings of new life in the sand and gravel of the river's bottom.

But they are being waylaid and netted by men in official green jackets and waders, standing in three feet of water. By an action called "stripping," biologists from the Department of Environmental Services massage female salmon over stainless steel bowls to extract their eggs. Periodically, a biologist will hold one of the males over a bowl and perform the same stripping massage, causing his milt to eject and fertilize the eggs. Once this stripping is completed, the salmon are returned to the river that will carry them back to the lake.

After some time, people—families with children and sometimes their dogs—drift back across the field to a lovely colonial home whose owners have graciously opened its doors for a community pot-luck dinner. And such a dinner it is! All manner of meats, casseroles, vegetables, desserts, and drinks are on hand in nearly limitless supply. Neighbors stand on the porch, plates in hand, talking and laughing. They sit inside on sofas and antique Windsor chairs, balancing plates on thighs while bringing each other up to date on the kids and jobs and life in general.

Others slowly drift about the three stories of the house admiring a collection of White Mountain Art, discreetly hung on every wall surface, that is possibly better than any display by a public museum. Paintings of primitive White Mountain scenes, composed in the latter half of the 1800s, are in the romantic genre of the Hudson River School. Many were painted in the area around Conway, NH. All in all, they portray a positive vision of the relationship between man and Earth. Lofty mountains with thundering cataracts and primeval forests elicit the wild purity of nature still untouched by the hand of man. The valleys, on the other hand, have been cleared and made into agricultural fields with fences, cattle, sheep, pigs, and chickens. Snug little homes sit peaceably along riverbanks, smoke gently curling from stone chimneys. People scatter about caring for the mundane chores of agrarian life—plowing, driving horses, mowing hay, bucking wood, culturing land and home and a life drawn from nature's pure and fertile bosom—all set amid a luminescence suggesting a divine human-Earth partnership.

I sense a resonance today, between the art on the walls and the scene before me of a pure river and lake and people participating in a run of fish that began long before the advent of man, either on this planet or around Lake Winnipesaukee. Native Abenakis took fish from this aquatic system for centuries before white men arrived. A little village on the other side of the lake, Weirs Beach, was so named because of the aboriginal fish traps that funneled fish into a narrow area where they could be netted for food. Although agriculture is largely gone now, several commercial truck farms operate today, providing a source of organic vegetables until after frost. In the larger picture, however, we continue as a community of people who value the land and waters and the good life that derives from their care. It is as though this wholesome community is a reflection of a wholesome ecosystem, still largely intact and little changed for thousands of years. The feeling that flows through this gathering of neighbors, the feeling that emerges from this rare collection of American art, the sense of integrity between lake and river and fish living together as they have for millennia—combines to amplify feelings of goodness, health, and hope for the future. An intact community reflects an intact ecosystem, and they combine to induce a positive feeling for life and the future of life gathered here together on the shores of "The Smile of the Great Spirit" (Abenaki meaning for "Winnipesaukee").

Self

We are all writers in some sense of the word. We spend our entire lives constructing a fiction, a narrative with a central actor whose name is ego, me, myself, and I—first person singular. All of the scenes—the dramas, the victories and failures, sins and sacrifices, hopes and dreams, the conversations, and the motivations—are built around this star performer. While we spend inordinate amounts of time and effort shoring up this myth of an eternal me, we invest as little thought as possible in the inescapable fact of our approaching death and return to Earth—to the humus that became human.

The true nature of "self" is informational feedback from a catalog of connections and relationships, each monitoring feelings related to the organism's probabilities of positive or negative opportunities for continuance.

In *Self Comes to Mind* (2010), Antonio Damasio likens the self to the conductor of an orchestra who, although he may appear to be directing the convergence of various instruments for a successful performance is, in reality, a product of the blended sounds of the musical score. The story does not derive from a self; self is a product of the story.

Damasio believes this fictional self serves the ultimate purpose of maintaining systemic homeostasis that determines long-term survival, and questions whether this humanly conscious self, "began as an aggregate of the inchoate wills of all the cells in our body ... ?"[1]

Information is causal for feelings and emotions. These are composites of "Qualia"—the nature of internal feedback from conscious subjective experience, an evaluation of our place in the broader context of connections to all the systems and subsystems with which we are implicated in the grand web of life on Earth. Feelings thus derived have tremendous causal power for our attitudes and behaviors, and they are no longer as mysterious as in former times. They are based on reasons, and those reasons can now be explained in terms of electro-chemical reactions, cognitive science, mathematical calculations, and past experience.

Architecture of Feeling and Emotion

A little boy sits on the edge of a dock at the edge of the sea, watching stars. Like all little boys, he's playing with a stick and begins to stir the blackened water. His eyes grow wide as the water swirls in waves of green light, like luminescent fairy dust. It is as though the Milky Way has inverted to cool, green stars rising from the deep. In that moment, he is infected with a hunger to know what it is that links the fairy dust beneath the surface with the stars beyond his grasp.

But it wasn't fairy dust, just bioluminescent bacteria talking to one another about filling the sea with their offspring.

These are eaten by zooplankton which are then eaten by small fish which are eaten by bigger fish, each one ingesting the bacteria that gorge themselves on the contents of the host's digestive tract. They are then excreted as bioluminescent feces which is again eaten by other consumers. All of this serves one simple purpose—to fill the entire ocean with their kind. Yet, the ocean keeps them in check so they can never stop striving.

The process that led these bacteria to emit light is known as "Quorum Sensing." It is normal for bacteria to release chemical messengers, known as auto-inducers, into the environment. Generally, however, concentrations are not sufficient to alter genetic expression for the production of luciferin. But when threshold concentrations are achieved, genetic switches flip to produce bioluminescence. The bacteria are talking to one another. A tipping point is reached when bacteria, in a nutritive rich environment, calculate that the energy spent to achieve bioluminescence will be compensated by increased food supplies in the gut of animals which are attracted to the visible display. No conscious purpose is at work, rather, those that behaved this way were selected by evolution to survive and reproduce. It's all in the algorithm that is behind these not so simple calculations—just a string of yes/no code refined over eons of time.

Schools of fish mass to avoid predation. The signaling architecture is nearly identical for them as it is in bacteria. An approaching predator is seen by only a few, who instantly change their behavior. Others near them perceive this change and react instantly. Reactions domino through the swarm to produce the rapidly changing motion and tight grouping. This action creates confusion in predators. Swarming protects both the group and individual. The swarm is one big sensing device with each member initiating action based on the action of the few individuals closest in its field of vision. A loose grouping of individuals becomes a unified mind using new information to calculate probable outcomes and make choices conducive to life.

At the same time, something strangely wondrous is at work here— an emergent mind, minding the entire system whose unchanging parameters are the extension of life and maintenance of systemic balance. Like a swinging pendulum continuously seeking the middle, life communicates and calculates to extend itself while maintaining integrity with its fulcrum.

Bayesian Probabilities

Believing that we are derivatives of the patterns and processes of nature, tested and verified across long eons of time, I suggest that human feeling and emotion are causal for many of our choices and actions and operate on the very same, although more sophisticated, basis as those of lower forms of life. Let's start by looking over the shoulder of an English clergyman as he hunches over some calculations in his study ...

The oil lamp burns late this evening in 1755 in the study of Mt. Sion Presbyterian Chapel, Kent, England. The Rev. Thomas Bayes is tinkering with a formula to calculate statistical probabilities. Like many clergy of his day, Bayes is a polymath with interests in theology, philosophy, math, and science. He will develop Bayes Theorem—a calculation of probabilities based not on frequency distribution, but on confidence in information compared with the mental maps that undergird our beliefs and behavior. Analyzing new knowledge in light of past experience, we synthesize probable (because nobody knows everything about anything) outcomes relative to new factors or a changing environment. Probability, then, is a logical forward model—a probabilistic map— derived from calculations based on new information and our current mental maps. The artificial intelligence in Google's self-driving cars uses Bayesian software patterned on neural logic.

Alan Turing broke the Enigma Code based on his "Computational Theory of Mind." Neurons are wired together as pattern recognizers. Knowledge and sensory data are analyzed and stored in these patterns or experiential maps. Recognition and inference cause logic gates to open, releasing sodium, potassium, and calcium ions, amino acids, peptides, and hormones into neurons to produce voltage potentials that fire information to other neurons. When a threshold number of neurons fire together, a quorum is achieved to initiate action, rejection, or confirmation. Decision is a probability calculation by consensus among the brain's neural pattern groups just as it is in bacteria, bees and ants, swarms of birds and fish, herds of caribou, and lions. From the most primitive cells to the human brain, nature decides based on computations involving new information, stored patterns, and quorum affirmation.

But the mind is more than a brain. For centuries, learned discourse has debated the "mind-body problem." In reality, there is no such thing. Mind and body are not separate entities, but a unity cycling information and feedback to maintain systemic balance.

Consider Margie Profet's "Morning Sickness Hypothesis": Many pregnant women become nauseated by certain foods. These foods may contain toxins, harmless to the mother, but potential sources of birth defects at the stage when embryonic organs are just forming and vulnerable. Generally, morning sickness wanes once embryonic organs/systems are fully formed. Women with severe morning sickness are less likely to miscarry, and their children have statistically fewer birth defects. The brain and body communicate as one, with feedback from the body interacting with stored evolutionary patterns, to foster and protect life. This is quorum-based thinking and deciding at a subconscious level because the mind and body are a unity displaying deep evolutionary memory built to react for long-term survival.

For 98% of our history, we lived as hunter-gatherers in small nomadic bands. We partook of the land's affordances, living in caves and wooden shelters, gathering wild foods, bringing down large animals by team effort, and making decisions based on a psycho-chemical architecture that began with bacteria. In simplest terms, we are in a continuum built on a program derived from every life form and experiences that went before. Quorum sensing and Bayesian probabilistic calculations are foundational to the way our brains and behavior works. This is often referred to as Bayesian Probabilistic Calculation or Bayesian hierarchical predictive coding.[2]

Psychologists understand the critical value of feelings and emotions in human behavior and have spent decades seeking their roots. This is important because many of the most determinative decisions we make are influenced as much by how we feel as they are by pure reason. With *Molecules of Emotion* (1997), Candace Pert documented the integration of body and mind in the role of feeling. As a graduate student, she did research with neurotransmitters and was the first to discover opiate receptor sights in the brain.

Neurotransmitters are proteins that affect the operation of neurons, causing them to secrete other chemicals, open various channels, activate receptor sights, and build voltage potentials that send information across synapses to activate behavior, confirm patterns, access or add to memory. These protein substances that bind to receptor sites are called "ligands," from the Latin *ligare*, "something that binds." According to Pert, "95 percent of (ligands) are peptides…these chemicals play a wide role in regulating practically all life processes and are indeed the other half of the equation of what I call the molecules of emotion." She references the work of the late Francis Schmitt who called peptides, "informational substances" because they function linguistically to distribute information about subjective feeling (qualia). Summarizing, she writes, "Neuropeptides and their receptors thus join the brain, glands, and immune system in a network of communication between brain and body, probably representing the biochemical substrate of emotion."[3]

Emotion is our weathervane, turned on its axis by the aggregate messages striking it at any one moment. Feelings are feedback sending coded information to the helmsman (*kubernesis*) by which he may make subtle or drastic changes in the coordination and control of the overall system based on instantaneous Bayesian probability calculations for the best outcomes. These signals often result in a "gut feeling" because the enteric nervous system contains millions of neurons that fire under the right conditions to cause the muscles of the abdomen to contract or relax. This portion of the nervous/peptide system is often referred to as the "the second brain." Neural impulses and chemical messengers form a syntax of coded signals just as words arranged syntactically are linguistically coded signals.

How do you feel right now? Why do you feel that way? Feelings are important, often more than facts. Feelings may even suppress facts. It seems that a whole lot of people don't feel well today. In 2016 there were

45,000 suicides in America and 64,000 deaths from drug overdose. One-third of Americans are taking medications that can cause depression. Meanwhile, we are getting fatter, families are fracturing, real income is declining, identity politics is taking over, alliances are unraveling, financial inequality is increasing, global warming is accelerating, and debt is rising 66% faster than gross domestic product. Feeling down? Synthesizing a broad range of concepts from cognitive science, biology, information theory, and physics, I am suggesting a new model—The Opportunity-Computational Hypothesis of Emotion.

Opportunity-Computational Hypothesis of Emotion

Biologically, life evolved to survive and successfully launch its DNA into the future. This we have in common with trees, toads, tigers, and tarantulas. To accomplish those goals, we must protect and nourish bodies and offspring. This requires access to resources (primarily energy) via physical health, networked social relations, education, material goods, personal, and financial growth. The human brain is capable of a quadrillion-calculations per second, and instantly computes potential for energy acquisition or loss and reproductive outcomes based on new information, using hierarchical predictive probabilities (Bayesian). New information arrives: You lost your job; you were accepted to MIT; that beautiful girl said, "Yes"; your car blew-up and you can't afford another; you're getting a 25% pay increase; the doctor said, "Melanoma"; the bank is foreclosing; you're planning to climb Denali. If personalized, each of these generates a feeling based on the brain's computation for future expansion or contraction of life opportunity and future energy/growth/reproductive probabilities (money is stored energy). Prolonged negative feeling sets the stage for binge-eating, depression, addictions, anger (fear), violence, social disintegration, and immunodeficiency diseases (depression suppresses the immune system).

Positive opportunistic probabilities, however, trend to optimism, healthy eating, exercise, strong family and social bonds, entrepreneurism, personal development through training and education, fewer health issues and medications, and lower rates of crime and addictions.

Beyond a certain point, money does not make people feel better. Solid families, inclusive relationships, fulfilled potentials, and growth opportunities contribute to positive feelings. Public policy that favors health care, housing, education and research, job-training, childcare, small business, a healthy environment, and public transportation induce better feelings. The current tenor of public policy that favors capital accumulation over creation of human opportunity suggests that future negative manifestations will only increase. The more optimistic 1950s, on the contrary, saw massive investments in health, education, housing, debt reduction, science, and infrastructure. These things, I suggest, were behind the relative euphoria of that decade, just as the disasters of racial strife, assassinations of popular leaders, and the Vietnam War were main sources for the social paroxysms of the 1960s. The current tenor (2020) of public policy favoring the elite/military/industrial complex at the expense of programs for human and environmental protection and enrichment carry heavy portents for future social fragmentation.

And the maddening thing is that catastrophe is often a conscious choice. Think back a few paragraphs to Bayes'Theorem and the mental maps we carry around. Ideally, when new information is received the brain modifies its maps for positive future probabilistic outcomes. Healthy mental process readily incorporates new information and uses that information for future guidance. However, it appears that many brains willfully dismiss new information that is contrary to their ideological indoctrinations—evolution is a hoax, global warming is a hoax, Covid-19 is a hoax, white people are superior to black people, etc. They refuse to change their mental maps based on empiric evidence and find themselves on dead end roads far from the new interstate highway. Feeling stuck and downwardly mobile, they begin to self-destruct or lash out at those who are building a newer, better way. This denial of new information and accompanying change is at the root of our culture wars, racism, homophobia, pollution, pandemic spread, and global warming.

Transcendent Experience

Harvard psychologist William James devoted a large portion of his professional career documenting *The Varieties of Religious Experience* (1902). James defined religion in broad, universalist tones rather than narrow partisan interpretations, "...religion...consists of the belief that there is an unseen order, and that our supreme good lies in harmoniously adjusting ourselves thereto."[4] The "unseen order" that James mentions corresponds, I believe, to the logically structured patterns that have stood the test of time to provide the scaffolding for a harmoniously ordered universal metasystem. James' ideas are close to those of German theologian Adolf Harnack (1851-1930), who wrote, "The great central fact in human life is the coming into a conscious vital realization of our oneness with this Infinite Life, and the opening of ourselves fully to this divine inflow."[5] Every experience recounted throughout James' book is simple commentary on Harnack's observation of coming into, "conscious vital realization of our oneness with this Infinite Life."

Ralph Waldo Emerson is most well known as the titular head of New England's Transcendentalist movement of the 1800s that emphasized man's unity with Nature. The Hudson River School, and other romantic painters of the same period, attempted to capture this transcendence—man's unity with benign nature—in paintings full of a luminescent grandeur, bountiful nature, and man's joyful relationship with an ordered, benign, and implied Providence. In so many ways, this rightful in-patterning of man within the unity of the natural world is an underlying theme in the American experience. Coming to America, European man felt that he had washed up on the shores of what was variously interpreted as both a waste howling wilderness ready to rend human presumptive arrogance, and the bosom of wild nature set here like the Garden of Eden waiting to offer her bounty to all who would lovingly nurture this soil. American's first true religion was naturalism.

Slowly guiding his motorboat to their camp on an island in Maine on a pitch-black summer's night, theoretical physicist Alan Lightman stopped to look into the eyes of the universe and ponder the depth of space, "And I felt myself falling into infinity...I felt a merging with something far larger than myself, a grand and eternal unity..."[6] Writing to Sigmund Freud, Romain Rolland (1915 Nobel Prize in Literature) linked religious sentiment with mystical experience—"a sensation of eternity...a feeling of something limitless...a feeling of an indissoluble bond, of being at one with the external world as a whole" (Letters).

A second avenue of transcendence lies in a multitude of near-death-experiences (NDE). NDE books have become a cottage industry, from Raymond Moody's *Life After Life,* to Elisabeth Kubler-Ross's *On Death and Dying.* An Amazon search for near death experience books yields titles by the dozens. Typically, a person feels that she has died, comes out of her body, sees an approaching bright light, looks down on the body that was her, and feels an overwhelming sense of love and unity with the universe.

Michael Pollan, in his book *How to Change Your Mind* (2018), notes two themes that recur over and over again among people using psychedelic drugs, specifically LSD, Psilocybin, MDMA, and 5-Meo-DMT (the toad). First, there is a near universal loss of all sense of "self"—to the point that the fear of death becomes almost laughable. And secondly, trippers invariably sense a unity with all of nature and the universe, such that they become one with the leaves of a tree or stars or the air gently caressing a pair of sheer curtains in an open window.

Here we have three different sources—three diverse types of human experience all pointing in the same direction. Individual religious experience has, for millennia, issued in a sense of leaving oneself behind in a journey that unites the seeker with the universe. Loss, or death of self, is a primary ingredient in the teachings of nearly all major world religions. This is followed by some form of union with the Greater Being of the Universe. Christians are united with Jesus by baptism; Aboriginals unite with the universe through a vision quest; Buddhists deny the reality of self in the first place and recognize that all are one. Transcendent vision has long been an experience of individuals enthralled by the power and continuance of nature.

Tennyson, Whitman, Wordsworth, Blake, Schiller, and artists of the Romantic Period all sought to convey the merging of individuality into the broader flowing river of time and nature. Lastly, mind-altering experiences under the influence of psychedelic drugs have been the experience of indigenous cultures for thousands of years. Native Mexicans were using magic mushrooms long before the Spaniards arrived. Mescaline and ayahuasca were tools of shamans prior to the burning of Rome.

And yet, if we look closely at these three broadly varied avenues of transcendence, we find nearly identical manifestations—a lost sense of self with all of its fears and vulnerabilities, and a merging with the being of a universe that is benign, if not loving, in its essence. By definition, therefore, transcendent experience is religious experience since the basic meaning of the word "religion" is *re-ligio* = to connect again. Religion is reconnecting with the source (more on this in the next chapter) ...

We once owned and operated a Christmas tree farm and nursery. Come December, we gathered truckloads of branches from overgrown and cull evergreens in our fields. These were brought into a large greenhouse that served both as a working display and sales center. Here, we cut branch tips and fashioned them into wreaths from 8" to 72" in diameter, along with garland, kissing balls, and swags. Filled with fresh wreaths, paperwhites, and piles of clippings, the greenhouse was an olfactory delight transporting people deep into Christmas-land.

I remember one particular late December afternoon. Customers had departed with their treasures and we were cleaning-up. Suddenly, the greenhouse transformed into a magical place. On this bright and bitterly cold December day temperatures had dropped, and the four, 48-light picture windows at the front of the greenhouse transformed into a fantasy garden. While the heat was turned up, the windows were constantly fogged and dripping with water. By late afternoon, however, with the thermostat turned down, water began to change phase and grew into frost ferns slowly creeping across the glass. For a few short minutes, as the sun stopped on the western ridge, some mysterious alchemy cast the ferns into a garden of golden paisley patterns, burnished in the last rays of a dying December sun. The air was filled with the scent of freshly cut balsam boughs and flowering paperwhites. Pachelbel's canon played enchantingly in the background. In that moment, I felt transported to an emergent reality orders of magnitude more elevated than that of common day ...

I believe transcendence emerges when we experience the convergence of multiple whole patterns, as I did on that day in December— beautiful ever-greened wreaths and garland hung about the walls, pure white flowers opened to full blossom, the air heavy with the smell of field and flower, deeply resonant music, and the last bright shafts of sunlight as they played among paisley-patterned frost ferns, gilding them into an exquisite and ephemeral work of art. The mind senses and recognizes this multi-patterned wholeness as a mark of "integrity." Whole patterns carry within them all the order and rightness of things that is the basis of truth. And if God is nothing else, God is truth. To experience truth is to experience divinity, even if but for a fleeting moment in time. Perhaps just one of these patterns, isolated, might not have had such a deeply moving effect. Combined, however, harmonic, multi-systemic convergence increases the amplitude of the brain's waveforms such that we enter an emergent state of "unitive consciousness." We become one with Aristotle's God—the eternal order of the universe. For a brief minute, we come out of ourselves and experience that scintilla of divinity that binds us to all other "forms" of divinity.

Feeling and Causation

Feelings are like the guard rails that keep us on the road of homeostasis—a state of balanced interaction within our social, physiological, economic, ecological, political, and cultural environments. Perhaps a good analogy would be the sensors in new vehicles that warn the driver when she is moving near the center or edge of the road. Driving our daughter's SUV last winter, I veered toward the road edge and felt the steering wheel begin to vibrate. The vibration was feedback from the sensors to alert me to danger and was causal for a conscious correction of path. I was in danger of leaving safe space. Correction was avoidance of harm based on instant Bayesian probabilistic calculation for the maintenance of homeostasis. This basic model functions in multiple areas of our lives. Reaction to a lump in the breast, a slowdown in factory orders, North Korea launches five test missiles, two years without a salary increase, increased average global temperatures, an erratic and unpredictable president of the US, constant wars, rising opioid addictions, tensions with your spouse—all cause sensed feelings that there is a threat to our future stability and well-being. They cause us to think and act in ways that maintain balance. Sometimes, however, there is nothing we can do for things beyond our control and we react by addictive behavior. We avoid or isolate the negative feedback hoping it will self-correct. If the negative feedback continues, it bodes ill for the individual and society.

Mental breakdown is the inability of mind to process information logically.

Often, however, it is not the reality of a threat but the intimation of threat that is causal for a certain course of behavior. Consider these statistics noted by Yuval N. Harari in his *21 Lessons for the 21st Century*: Average deaths from (foreign) terrorists post 9/11 were 50 in the European Union, ten in the US, and seven in China. During the same period, traffic fatalities in those respective places were 80,000, 40,000, and 270,000. Meanwhile, diabetes and excess sugar killed nearly 3.5 million, and air pollution took another seven million/people/year.[7]

Harari's point, if I'm correct, is that the threat of terrorism has been an effective tool to manipulate public feeling for political purposes. The threat of terrorism and some conjured weapons of mass destruction were tools for selling the Iraqi and Afghanistan wars. The threat of terrorism gave the government sweeping new powers of surveillance on American citizens. All of these manufactured threats avoided the true story: Saudi Arabia was home to Osama bin Laden, fifteen of the nineteen 9/11 terrorists, and is the main financial supporter of radical Islamic Wahhabism. But there was never a mention of invading Saudi Arabia. Moreover, the Trump administration wants to give them advanced nuclear technology and $100 billion of military weapons. At the same time, America is purchasing $60 billion worth of their oil every day of the year (current numbers =.875 m/barrels/day X $72/ barrel). In effect, we are giving financial and military support to the very source of the terrorism we are allegedly fighting.

The central reason for ill-feeling in so many millions of people today is a corrupted message developed by leaders to induce mass fear in order to manipulate the political process so they can make more money. Ironically, they are bankrupting the nation in the process. Osama bin Laden spent very little money to accomplish his goals on 9/11. In response, America has spent upwards of $5-$7 trillion (one-third of US national debt, now on our grandchildrens' credit cards) to fight never ending wars that have accomplished nothing but to deplete our finances and credibility around the world. Bin Laden's strategy, as judged by harm and expense to the opponent, was and continues to be vastly superior to that of the American military. As Harari observed, bin Laden was the little fly that buzzed in the bull's ear and caused him to wreck the china shop.[8]

The link between feeling and causality cannot be overstated. Drug addictions, suicides, PTSD, divorces, rising inequity, endless wars, ecological collapse, extinctions, rising national debt must be seen as honest systemic feedback to tell us that we have a systemic problem that will not fix itself.

If, therefore, the patterns of mind and patterns of Earth are mirror images, then distortion of pattern in one sets up resonant conflict in the other. A mind subject to the destructive patterns senses and internalizes conflicted illogic.

Any constructive path going forward must cultivate a positive emotional response. I will contend, in a later chapter, that feelings of belonging, or union with the greater cosmos are the best motivators for constructive change in our relationships with one another and with the Earth. The question, "What are we against?" will bring only division, pain, and suffering. Moreover, we must hold front and center the question, "What do we love?" Love fosters harmony, caring, commitment, and sacrifice on behalf of the beloved. Consider these words of Sigmund Freud, "A man who is in love declares that 'I' and 'you' are one and is prepared to behave as if it were a fact."[9]

Feelings are causal, and until we feel our common union with each another, the broader environmentally destructive patterns of behavior will continue until both human and Earth systems are driven to dementia.

11. Social Institutions

"This aggressive human monoculture has proven astoundingly
virulent but also toxic, cannibalistic, and self-destructive."
— Roy Scranton[1]
"When a network is already converged to a specific set of patterns
and partially pruned, it will require many more repetitions to
'overlearn' a new—especially a contradictory—pattern."
— Jeffrey Satinover[2]

Religion

SLIP ON A PAIR OF SNEAKERS and let's take a walk in the woods.
Feel that soft, spongy texture from centuries of decayed leaves and
rootlets that cushions and welcomes your feet? The tactile sense is just
another way, along with the damp earthy smells, birds calling, and
light splashing across boles and boulders, that we waken to a living
presence that our DNA has known for eons of time. Connection is at
work here, for the forest is also aware of us. Each step sends signals
to other life forms about us. Our very breaths waft upward, swirling
around leaves and into a million stomata that feed carbon dioxide into
the leaves' glucose machines. Oxygen, the plant's waste product, is for
us a necessity—plants and people on opposite ends of the teeter-totter.

The more scientists study forests, the more connections they find. We now know that every square meter of undisturbed forest soil contains about 12,000 miles of mycorrhizae (fungal filaments), and that these (along with masses of bacteria and roots) form an underground network of communication between every member of the forest. Ecologist Suzanne Simard (PhD thesis work) injected trees with radioactive carbon isotopes and discovered that carbon products are shared widely within a forest via networks of mycorrhizal fungus at long distances from the original injections.

When caterpillars begin munching on its leaves, the tree manufactures a natural insect repellent and sends messages via electrical impulses and pheromones through its roots, leaves, and symbiotic fungi to the plants around it. Receiving the message, they begin to preemptively manufacture their own repellent. Tree roots feed fungi and bacteria that cannot make their own food. In turn, the fungi and bacteria break down stone and organic matter to make nutrients available for the tree. Bacteria, through quorum sensing, form biofilms around small roots to protect, heal, and keep them moist. Trees become aware of their members that are sick or struggling and share food and water with them via this Wood-Wide-Web. (see Peter Wohlleben, *The Hidden Life of Trees*, 2015)

The natural world, therefore, is more than a collection of separate entities united by interaction—it is a network of woven strands of reciprocal relationships. It is community in the truest sense of the word because it holds life in common. What is good for one, is good for all. What harms one, harms all. Each individual in a community is linked by multiple connections to every other individual through a vast interlocking scaffolding that supports shared life. Connections are the spiritual (unseen) links that form a binding unity. Spirituality is connection. To break connection—to separate, to shun—is to disrupt spirit. To sever spirit is to collapse the scaffolding. A world of broken connections becomes dystopia, where all who formerly stood together, fall together. By definition, the religion of the universe is a unitive "Connectionism."

Let's again go back to basic definitions for understanding. Ligaments are connective tissues that bind bone to bone at joints. Religion, in its basic sense, reunites or reconnects disparate pieces and parts into a functional unity. Without the ability to move, facilitated by articulating joints united by ligaments, most higher organisms would perish for lack of resources. Lignin is the complex organic polymer that glues cells together and provides rigidity to plants so they can stand erect and gather sunlight. Therefore, the simplest and most primitive understanding of religion is that of a binding force that unites diverse parts into a mutually beneficial unity for increased survival of the whole.

On the other hand, we are the chief disrupters of natural networks. Modern man has become an extinction machine of the highest order. We specialize in broken connections, isolation, and dystrophy. For those who will see, examples are on every hand. Therefore, the only logical conclusion is that man has become a destroyer of the connections upon which life is built. Our arrogant ignorance is a blind and narcissistic delusion based on divisive religious fictions that have become, de facto, **de-ligions.**

As Cosmides and Tooby see it, "The cultural and social elements that mold the individual precede the individual and are external to the individual. The mind did not create them; they created the mind."[3] Across the eons of evolution, group living based on mutualism enhanced both individual and species survival. Therefore, mental patterns evolved that suppressed pure selfish purpose in favor of group welfare. Selfish mind can be an evolutionary liability. Conscious linear selfish purpose is disastrous in a world evolved on a platform of circular non-conscious causality regulated by distributed feedback to maintain homeostasis. How far back did this begin?

Primatologist Frans de Waal in *The Bonobo and the Atheist* (2013) cites multiple examples of not only cooperative but altruistic behavior in our ancestors. Time after time, he observed chimps and bonobos helping each other, grooming, sharing food, assisting the old and sick, patching social relations after a fight, rebelling over unfair treatment of another member, and punishing members that hurt the group cause. He refers to Jane Goodall's conclusion that chimps exhibit religious behavior in their waterfall dance that may be reflective of certain religious ceremonies among humans.[4]

These examples serve to support de Waal's thesis that morality and ethics—the basis of religion—originated bottom up rather than top down. The roots of religion lie in the social compact based on equity, fair play, empathy, egalitarianism, and basic justice that are indispensable for group survival and continuance. Morality is an evolutionarily selected code of behavior that arose long before the Code of Hammurabi. Human religion is simply an amplified and ritualized modification of something that has been around for billions of years. But this is only part of the explanation. The roots of religion have an even deeper source.

I believe that religion, as expressed by cooperation to enhance overall survival and reproductive success, is a response to the thermodynamic necessities of concentrating energy in an entropic universe where the natural tendency is to randomness and disintegration. Behavior that favors energy concentration, favors survival and rising complexity, and was selected. Bacteria, bees, and bonobos do it. Religious behavior was molded by the laws of thermodynamics.

In a 2019 article for the BBC, Brandon Ambrosino addressed the question, "How and why did religion evolve?" (BBC 4/27/19). He begins with the history of food sharing and the sacral meal, then proceeds to the rise of awe as a religious response to that which is beyond us and evolved in shared rituals that unite people into a community. Shared identity as a community facilitated cooperative behavior which increased the probability of survival and reproduction. Much of religion is based on feeling, which I would call the "social lignin" that glues a community together as much as auto-inducers bind a quorum of bacteria into a thin film. The four primary emotions are aggression, fear, sadness, and happiness. Three are negative, and one is positive. And this is where religious instinct, which is a positive necessity for human survival, can be manipulated into a tool of control, repression, and the downright demonic malevolence of every imaginable variety that is currently threatening the entire global order.

Ask any person who has disengaged from a former religion about his or her motivations and you will get nearly identical responses to the effect of: "It was all based on guilt and control," or "they were hypocrites teaching one thing and doing just the opposite," or "to be happy in that religion you had to check your brain at the door and stop thinking for yourself," or "I felt manipulated to conform to something that increasingly made less sense," or "it was just a ruse for political power and influence that had nothing to do with love and compassion." The upshot is that religion, whose motive should be to foster an enhanced society of connected individuals, often becomes a tool for division to achieve the political and economic goals of the ruling elite. This was the case in the time of Jesus, and it continues today. (See Reza Aslan's *Zealot* [2014] for historic documentation.)

Dominionism, derived from the assumed superiority of a religion and the right of its adherents to control Earth and others for their ends, blasted onto the scene with the advent of the Abrahamic religions. Judaism, Islam, and Christianity are revealed religions, deriving their authority from religious texts allegedly revealed by God Almighty to men he chose to be the vessels of his word and will. These texts, of necessity, must be divinely inspired for that religion to gain authority over all competing ideologies. Therefore, if Genesis says God created everything in a six-day marathon, that's the way it was. Regardless of any and all evidence, religious authorities demand that the Biblical record is the one sure source of truth and therefore the theory of evolution must be a Satanic ruse to draw people away from God-given truth.

Look back at those four primary emotions. Communications experts have learned how to mold the message to tap into those emotions. The priest has vast power over people because he is God's ordained vessel to communicate divine will. He can use fear, guilt, shame, and anger to manipulate human feelings and behavior. Religious leaders are invaluable to the military/industrial complex in making the case for war against "the enemies of God." Voltaire was right, "Those who can make you believe absurdities, can make you commit atrocities" (Questions sur les miracles, 1765).

Using the metaphor of prostitution, John, in his Apocalypse, sees in the future a church that is more accommodative of commercial and political interests than it is for the message of love, reconciliation, selfless giving, equality, compassion, and lifting the downtrodden in the name of Jesus. It was a future church that had sold its soul for power, aggrandizement, health, wealth, and prosperity. The impostors love their jets, expensive suits, and golf resorts. They will not speak truth to power, because power, not God, gives them what they want.

Although many of the elite have no personal use for religion, they will defend it to the death because it is the cheapest and most effective way to keep a trusting populace in their harnesses. What facts support the notion of a "chosen people"? What empiric evidence decrees the divine right of kings? It's all a ruse for the elite to hold power. They need their archbishops, mullahs, temples, cathedrals, and gold vestments to keep the masses between fear and hope and prop up the entire charade amid a grand pageantry.

How ironic that religion which was intended to be a force of healing, compassion, love, and unity has become a driver of division, war, and hatred in the western world today. It exists as a contradiction. Two thousand years of Christianity should have produced a utopian society instead of the vicious partisanship and division we are experiencing today, that is more willing to kill those who disagree than to offer a cup of water in Christ's name. The Abrahamic religions have killed more people in the name of God than all the followers of Baal and Huitzilopochtli. Religious wolves, hiding under sheep's clothing, promote death and division rather than peace. What was intended to *re-ligio*, has become *de-ligio*, which, for a few, is more financially profitable.

Culture

The simplest definition of culture is, "the social behavior and norms found in a society" (Wikipedia). I feel most comfortable (from my horticultural background) associating the word "culture" with "cultivate." To cultivate is to employ the best techniques for growth and increase. Positive aspects of cultivation are nourishment, optimal environmental conditions, guidance, water management, and exposure to light. Negative aspects focus on disease prevention through healthy media, protection from opportunistic insects, animals, and environmental forces such as wind and hail, and excessive competition from weedy species.

Translating these motivations into the human sphere is reasonably literal. Culture is best when individuals and groups have optimal conditions for growth and development. These include, but are not limited to, balanced nutrition, a clean environment free of toxins, adequate shelter, protection from environmental hazards, protection from opportunists who would parasitize and rob, education (including the best in the arts, sciences, and humanities), and an overall encouragement to develop beyond the last generation. The immediate goal of culture is healthy development. The ultimate goal of culture is transcendence of former achievements that issue in more elevated living. Therefore, culture and evolution are teleological processes concerned with maximum fulfillment of potential—Aristotle's avenue to higher being.

Culture advances as multi-disciplinary knowledge and exposure spreads via sharing among individuals and groups. The greater the number of persons who can share in broader pools of knowledge, the more cross fertilization will occur, which will result in increasingly creative solutions for future progress and problem solving. Businesses have learned the benefit of "skunkworks teams"—small groups of diverse discipline individuals called together to come up with solutions to knotty problems. A team composed of a physicist, linguist, mathematician, biologist, and an engineer will have a multifaceted point of reference that could offer more novel approaches than a team of five engineers. We all have a tendency to think within our own disciplinary boxes and can benefit greatly from other points of view. Cultural echo chambers, where we hear only our own doctrines, are like populations of animals that continually in-breed within the family. Eventually, their more homogenous outlook becomes inadequate for dealing with a heterogeneous world.

This is borne out by history. Culture always advances faster in areas where diverse groups of people interact for mutual benefit. On a trip to China several years ago, I entered upon the cobbled streets of old Lijiang, founded 800 years ago on the ramparts of the Himalayan Mountains. Old Town Lijiang was built along the Yuhe River, dancing fresh and pure from the mountains above. Ancient engineers divided the waters through a manifold that sends a stream of water flowing through every concourse. This provided fresh water for all the homes and businesses along the streets, as well as a place to bathe and empty chamber pots.

Lijiang was one of the eastern ends of the Old Tea Horse Caravan Trail, thus people have had social intercourse with middle easterners as well as Europeans for centuries. As a result, the old town is rich with culture and an absolute delight to the senses. The foods, music, art, languages, and history are rich, satisfying, and deep. If I could build, in America, a pattern city for the enhancement of human life and exchange, it would be modeled on Old Town Lijiang, using the best of modern technology and sustainable energy, with pedestrian only streets, flowing waters, massive gardens, and al fresco dining. Minds generate emergent concepts from cross-fertilization.

Susan Blackmore states her thesis at the outset of *The Meme Machine*: "[What] makes us different is our ability to imitate."[5] The word "meme" is a shortening of the Greek, *mimema* = imitating, to imitate. Richard Dawkins coined the word *meme*, proposing it as a unit of cultural evolution much the same as "gene" is the unit of biological evolution. The Oxford English Dictionary defines "meme" as, "An element of a culture that may be considered to be passed on by non-genetic means, esp. imitation."

Genes are coded algorithms for constructing proteins. By the recombinatory process, genes exchange code with other genes, or their expression is altered by epigenetic factors, such that novel codes and combinations arise over time, prove more efficient, and gradually replace older and less efficient codes. Genetic code is biological information that is replicated and passed on to future generations. Memetic code is information carried in the cultural patterns of the brain that determines ethics, morality, and social behavior. These cultural patterns are handed down across multiple generations, modified somewhat by each new generation, and dictate understandings of ourselves, society, social institutions, ideological orthodoxy, and relational behavior. Memes that group naturally and reinforce one another are called *memeplexes*. Blackmore, Dawkins, and Dennett all agree that religion is a major memeplex—a cluster of memes that heavily influences one's worldview—that becomes a controlling cultural paradigm. The larger memeplex, however, is the one involving the illusory concept of a self. Blackmore considers this false sense of a separate self our greatest memeplex.

However, like genes, memes can carry erroneous and potentially deadly ideas that can result in the death of both the carrier and his culture. In other words, like malware used to infect computer code, virulent memes can be inserted into a culture by way of communication processes that will, at some later time, become destructive. Computer malware often comes with the offer of something really good ... "you won the lottery, just click here ... send $1000 today and we will release monies held just for you...verify your password and we will release funds" ... They suggest life-expansion, like Mephistopheles offering the sweet, pure Margaret to Faust. Asked by Faust about his true nature, Mephistopheles replied, "A part of that force which always seeks evil and always does good."

This is the double-edged sword of advanced technologies—they promise good but can deliver horrendous evil. Wiener, Einstein, and Oppenheimer were highly ambivalent regarding the results of scientific and mathematical advances which paved the way for atomic and thermonuclear weapons that contain enough power to wipe out entire civilizations, if not life on Earth. Thus, Oppenheimer's quote from the Bhagavat Ghita, "I am become death..." Social networks have become exceptionally popular but are often used as avenues for memetic infections and ideological manipulation.

Memes seek reproductive advantage. If successful, they spread to the wider culture and are passed on. Religion, culture, the illusion of a separate self, economic, and political memes are selected but may be false code that decreases reproductive advantage. The memetic "good life" of carbon-fueled capitalist consumerism has given us a lifestyle beyond the wildest dreams of our ancestors. Unfortunately, scientific data suggest that consumptive lifestyles powered by fossil fuels may also result in a Faustian climax that ends in severe regression.

The major motivation for a new Kuhnsian paradigm will come when we see that the old memes have become anomalies threatening survival and reproduction. Cultural patterns of destructive industrial practice, based on a throw-away consumerist mentality, are showing negative biological value. Religious/cultural memes often encouraged increased births. Meanwhile, social balance is teetering in the face of overpopulation, fewer per capita resources, and massive migration from third world nations that threatens the cultural survival of advanced nations unwilling to support them. Reproductive advantage is pushing much of the world away from the dinner table.

Homo sapiens—man the wise—is a teller of stories. The cave paintings of Lascaux and Chauvet of deer and buffalo dripping red ochre as they dance mythically in the flickering light of a torch, hand prints chipped on a sandstone cliff above a slithering serpent in the desert southwest, Odysseus sailing between the twin perils of Scylla and Charybdis, God speaking from a burning bush on Mt. Sinai, the infallible, invisible hand of the market, a new world coming down from heaven, Spider Woman and her sipapu, a chosen people, dualism, dominionism, and death—all are cultural/religious myths that help us understand our place in the world.

Stories are metaphors that give rise to culture. In the end I—this person called "myself"—am just a story, a fiction built of experience and sensation woven into a narrative with "myself" as the central character who charges across the stage of life like a red ochred dream dashing across the wall of the cave, until a passing wind extinguishes the torch's flame.

We identify with stories because we are a story. So long as our fictions cohere in a smoothly flowing narrative, all is well. But when they decohere, unravel, become conflicted, they become harmful. We are currently in the midst of a period of great decoherence between new understandings of reality and the old cultural narratives. These cultural stories, many now seriously corrupted, are in conflict with reality: white male patriarchy, unregulated capitalism, democratic governance, divine creation, the authenticity and inspiration of religious texts, racial superiority, dualism, dominionism, death, a chosen people, and manifest destiny. None of the above list is a separate entity unto itself. All are interconnected in a memetic web that is depatterning the natural order of Earth.

Because of the cultural myths of racial superiority, this nation is still fighting a low-grade civil war. The current administration uses the terms, "thugs," "criminals," "rapists," and "animals" to describe immigrants seeking asylum in America. This is a conscious attempt to portray these people as inferior, a threat, a scourge to be resisted, if not murdered. It's the same language used by the Third Reich to justify genocide. It's the type of language we used to provide cover for our genocide and displacement of Native Americans.

Threats to cultural stories are interpreted as threats to individual stories because culture provides a large portion of our identities. By identity protective cognition and cognitive dissonance, people lie, deny, rationalize, or seek to discredit honest sources in order to maintain the integrity of their story. We believe one thing and do another. German doctors and guards, members of good standing in the Church, could go to their assigned concentration camps, carry out all manner of inhuman experiments or kill Jews by the hundreds, and return to church and receive holy communion. People will resort to all forms of dishonesty, confirmation bias, and rationalization to protect their cultural story and the ego built upon that scaffolding, even in the face of overwhelming evidence that their story is demonstrably false.

Cultural beliefs and practices derived from the "religions of the book" are proving toxic to life and, with continued support, may erase the very lives they claim to ennoble. Cultural memes are threatening the survival of biological genes. Unless certain cultural memes are replaced by more sustainable, inclusive, and long-range paradigms, the future of civilization could be bleak. Memes should exist for the fullest development of genes, not vice versa.

Politics

In his *Republic*, Plato counsels that democracy is the poorest form of government. It is destined to end in tyranny, he says, because the Sophists (the elite) are quite competent at manipulating the hoi polloi (the masses). Those who can manipulate the message in their favor will ultimately gain control. Never has this been truer than today. The elite have at their disposal the latest psychological research, unlimited funding, control much of the media and politicians by way of campaign finance and distributed sugar plums, and advertising agencies that craft the message with great subtlety to appeal to the reptilian brain in our brainstem and manipulate primitive emotions in such a way that people will vote based on stoked fears and insecurities. Modern political campaigns are largely about pitting groups against groups with the express purpose of driving wedges to heighten stress and move voters toward candidates who seem to feel their pain and are offering short term placation. Candidates are portrayed as saviors who will put "those people" in their place and restore culture to its "glory days"—"Make America Great Again."

Conflict and manufactured crises work to the advantage of those who seek power and control. While they offer short-term and simplistic bromides, their long-term plan of dividing people to win elections so they can strip assets sows such instabilities within the system that it may ultimately collapse into chaos.

Early on, Norbert Wiener and his Macy Conference cyberneticians discovered the overwhelming power of message for systemic command and control. Among men, information is distilled into language and passed by communication. Douglas Hofstadter wrote, "It is through language most of all that our brains can exert a fair measure of indirect control over other humans' bodies...My brain is attached to your body via channels of communication...I can also control your body indirectly."[6]

Once the fact of anthropogenic global warming was established as a scientific fact (Exxon Mobil understood this in the 1970s), the fossil fuel industries began crafting a media message to instill doubt and confusion, while cultivating think tank bias and favorable politicians through directed funding and contributions. The highly crafted message portrayed global warming as an overblown media story, a hoax propagated by liberals, and an expression of either new age religion or Neo-Marxist propaganda to destroy free-market capitalism.[7] The effect was confusion that paralyzed decision making.

Fearing for their profits, automotive manufacturers and fossil fuel corporations, along with the National Association of Manufacturers and Chamber of Commerce formed the Global Climate Coalition. Together, they produced a film entitled *The Greening of Planet Earth*.

The message was that doubling atmospheric carbon dioxide would enhance photosynthesis thereby making a greener, more agriculturally productive world with more resources for all. The film was given to national politicians and heavily influenced their movement away from the science of global warming. In the process, they hired two "scientists"—Fred Seitz and Fred Singer—who had both been involved in a program of propaganda for the $50 billion/year tobacco industry that had long understood the causal link between nicotine and addiction. All of this provided a political rationale to exit the Kyoto Protocol, based on the supposed conflicts within science, and potential damage to the economy.[8]

Later on, their weapon of corrupted communications would be useful in two ways: It would ridicule Al Gore's alarm about fossil fuel's contribution to rising global temperatures, thus questioning his presidential qualifications, and create broader suspicion that climate change was an attempt by the United Nations to create a "One World Government" based on Agenda 21 which proposed goals for sustainable development. Fossil fuel interests were able to craft a new movement in the Tea Party to oppose this imagined monolithic government that was out to take away people's freedoms and tax them heavily to support renewable energy.[9] These manipulations aligned people to the right in a newly solidified Republican Party that had successfully raised the specter of a liberal/Communist conspiracy intent on world domination.

Manipulated by the message of fear, huge grassroots organizational efforts by the likes of the Koch Brothers, and favorable legislation submitted by the American Legislative Exchange Council, right wing and ultraconservative politicians came to dominate local and state governments and squash efforts to advance sustainable energy. Ironically, while the message was freedom, they managed to keep us in bondage to jihadist Middle Eastern governments who largely fund their terrorist activities and education by American money spent on oil.

The largest effort of modern politics is manipulation of messages to gain control over "democratic" voters, to legitimize politicians who will enact policies favorable to the elite while harming the very voters who elected them. Once they gain access to our brains by triggering fear and division in the reptilian brain, and manufacture crises to reinforce their message, they can download their "message malware" into our brains and influence the vote.

The truly maddening thing about current political messaging is that any reasonably rational person can clearly see that the perpetrators of this malware do not believe their own message, just as those who pander to the religious right are quite often irreligious in their personal lives. Thomas Paine wrote of them in *The Age of Reason*: "Infidelity does not consist in believing, or in disbelieving; it consists in professing to believe what one does not believe. It is impossible to calculate the moral mischief ... that mental lying has produced in society. When a man has so far corrupted and prostituted the chastity of his mind, as to subscribe his professional belief to things he does not believe, he has prepared himself for the commission of every other crime."

But this is very profitable for the elite. They have had their taxes reduced and their assets are increasing (defense stocks continue to appreciate nicely). Meanwhile, lower and middle-income people have a smaller net worth than 20 years ago, the majority of new jobs are temporary and low paying, and the social safety net is being weakened in favor of sugar plums for the elite who control governance by controlling the message. But this charade can only continue so long until unheeded feedback (human and natural) builds beyond containment. Politicians would do well to understand the biological roots of governance.

Governance exists to ensure cooperative behavior that will benefit survival and reproduction. Morality arose to ensure equitable treatment of group members who cooperate better if they are treated fairly. Fairness and equality are basic to group survival. Refer to previous observations by primatologist Frans de Waal about fair play, cooperative behavior, unselfish sharing, peacemaking, and group discipline for errant conduct in tribes of bonobos and chimps.[10] Basic moral behavior and ethical treatment are hallmarks of interaction in all group animals (perhaps forests, too).[11] The roots of both government and religion predate man by billions of years.

According to the Preamble of our Constitution, government has 6-purposes: to insure a more perfect union, establish Justice, ensure domestic Tranquility, provide for the common defense, promote the general Welfare, and secure the Blessings of Liberty to ourselves and our posterity. Analyze these closely and you will see that they do not differ one whit from the unwritten purposes of an ant colony.

What is the ultimate purpose of government? Government must exist for the advancement of human potential, protection from threat, and the greatest good for the greatest number. No biological community can long hold together amid vastly inequitable treatment and opportunity for the masses of its constituents. Government must be "of the people, by the people, and for the people." This binge of selfish aggrandizement cannot last long. Democracy is on the road to ruin at the hands of its own democracy manipulating officials. Unless big and dark monies are removed from electoral campaigns and replaced by honest debate over pressing issues, tyranny seems a foregone conclusion.

Economics

Ecology (study of the household) cannot be separated from economics (law of the household). Both disciplines concern the household as it benefits those who dwell in it. A house can neither stand if the supporting pillars are compromised, nor continue if its requirements exceed its resources. In both cases, the household will experience a crisis—in the first case from inadequate structural support, or the second from unrealistic and imbalanced management of resources. Ecology and economics are two sides of the same coin. Yet, what seems a simple, rational observation is rejected by the majority of reigning economic opinion. This is the root of the complaint by former World Bank economist Herman Daly.

The main thesis of Daly's *Beyond Growth: The Economics of Sustainable Development* (1996) is that the economy is a subsystem of the environment—a healthy economy depends on healthy ecosystems—ecosystemic deterioration will insure economic deterioration. His vision is one of a steady state economy that operates within boundaries that ensure a healthy ecosystem going forward. A steady state economy will ultimately be a zero-throughput economy, where input and output are balanced with no residual throughput, or waste garbage as is the case with pollution today.[12]

In practice, this means the elimination of fossil fuels which leave a throughput of carbon dioxide and heavy metals that issue in both climate warming and systemic illness. It means the elimination of garbage dumps and landfills in favor of 100% recycling of all vessels and containers, and the composting of all organics. It means mitigation where we must have some throughput, such as planting more land to trees in cases where coal or petroleum is our only option. It means more distributed food and energy production by placing production nearer consumers. It means electrifying transportation with 100% electric cars and high-speed rail, coast to coast, while working to either eliminate air travel or to invent electric-powered aircraft.

According to Daly, economic globalization runs counter to the spirit of free-market capitalism espoused by Adam Smith. Smith's overall vision was one of communal benefit based on personal entrepreneurism. If individuals were permitted to take risks to build better mousetraps, employing both their genius and capital (or that of investors), they had a right to the free exercise of their business, and resulting rewards (if they do no harm), without government interference. Thus, the selfish motive for financial gain should not be thwarted, nor should profits be unduly siphoned off in such a manner as to discourage future investment.

It would seem a common understanding that economics, as well as politics and religion, exists for the benefit of mankind. After all, what is money for? Is money not simply a medium of exchange whereby persons can access resources necessary for life and well-being? For what good reason should any member of a society be kept from quality health care and education? Especially, in the richest nation on Earth with trillions of dollars to dump on wars that, to date, have yielded no appreciable benefit?

Big business, the media, and the advertising business, in the largest single psychological manipulation in all of history, have created a fantasy world where every man and woman can be master of a domain where their every wish is fulfilled with only the slightest effort. Consumption makes us into little gods with the power to manipulate the resources of Earth (and a teeming mass of poor laborers who receive barely enough to survive) for the fulfillment of our least whim.

Due to the impetus of World War II, America became a manufacturing powerhouse. We made tanks and cloths and weapons and planes and all manner of supplies, along with supply chains to deliver resources from the far corners of the world with record speed. At the end of the war, however, big business worried over how to continue this massive cash cow of manufacturing once huge war-time demand ceased? The short answer, and the one China has embraced more recently, is that we must turn domestic populations into consumption machines.

It was retail analyst Victor Lebow who, as much as any one man, summarized the indoctrination required to convert America to this new religious *Consumerism* that would keep the factory lines running full speed while heaping unprecedented profits into the coffers of investors and corporations. "Our enormously productive economy … demands that we make consumption our way of life, that we convert the buying and use of goods into rituals, that we seek our spiritual satisfaction, our ego satisfaction in consumption…We need things consumed, burned up, worn out, replaced, and discarded at an ever increasing rate."[13]

Like Mephistopheles, Lebow failed to tell the American people that this would all come at a mortal price—a price we are just beginning, in these last five decades, to pay at a compounding rate of interest.

Modern industry operates largely on economic models of extraction and depletion. We take what we want, dump our garbage, and move on to another place or people ripe for extraction and depletion. The problem now is that ever-larger sections of Earth and mankind have been worked over, depleted, and left to fester in sickness and poverty. They are hotbeds of terrorism that have become a people's last hope to gain equity in this world.

What will we do when, by the advent of artificial intelligence, human labor becomes a liability rather than an asset to exploit, when we become a useless resource? Who will feed these people? Will the rich and powerful willingly submit to increased taxes to provide a living income for people no longer needed to power their industrial and war machines?

Engineered For Failure—Leveraging Darkness

The main point of systems theory is relatively simple: Message is primary. Message, either by external inputs or systemic feedback, controls and commands. The governor, the tiller, the steering wheel, the switch, the computer, the Federal Reserve, Google Inc., the Roman Catholic Church, the Pentagon, the Senate, the kidneys, a robin, an earthquake, rising global temperatures—all are agents controlled by message. A message can be delivered by the signaling flip of a wing, a wink, the press on an index finger, the slap of a beaver tail on water, Morse Code, smoke, moonrise, lengthening days in spring, a touch on the leg, an angry look, or a smile. Messages are signals, and signals are information. Our stories are messages by which truth is evaluated.

Dualism, free-market capitalism, the self and soul, heaven and hell, the divine right of rulers, dominionism—all are stories by which people guide their lives and organize societies. What we make of ourselves and this world is largely the result of the stories we live by. Since our stories have become accepted truth, we largely attempt to live in a manner that conforms to the story. If our stories really are true, we should expect beneficial outcomes. If they are not, or our lives do not conform to story, we might reasonably expect negative outcomes. Yet, if everyone believes and lives by the same story, should not outcomes be universally good? How can we really know if a story, especially one that has lasted thousands of years, is true? What is the basis of truth and how can we know it?

The main tool behind most of the systemic damage to Earth and life systems is the leveraging of dark areas. When messages are blocked or corrupted, the systemic mind breaks down and cannot maintain homeostasis.

The major religions of the West were delivered in darkness. God is removed from Earth in an unintelligible heaven. Religious texts were handed down to chosen authors (male) and codified based on word of mouth (subjective) testimony. The creation of man and the universe was by divine fiat, without the possibility of knowing process.

The world financial system of today lurks in the darkness of shell corporations, off-shore banking sheltered from investigation, back-room deals between regulators, corporations, and ratings agencies, and regulatory agencies run by industry insiders.

Politics is shrouded in the darkness of massive sums of monies delivered to political campaigns with no accounting of sources and protected from taxes via 501 exempt entities. Legislation is written by shadow groups, given to bought politicians, and voted into laws that overwhelmingly benefit special interests. Even though major voter feedback favors legislation that benefits the masses, politicians have so prostituted themselves to big monies that they ignore the will of their constituents. Wars are initiated based on unverified messaging.

Capitalist Crises

William Robinson, professor of sociology at University of California, Santa Barbara, wrote an article in the April 2, 2019 issue of *Science and Society* titled, "The Global Capitalist Crisis and 21st Century Fascism: Beyond the Trump Hype." Briefly, his thesis is that fascism is a response to a crisis within capitalism and compares the current movement to nationalist/populist politics with the rise of fascism during the Great Depression of the 20th century in Nazi Germany. The root of these crises, he proposes, is the lost legitimacy (trust) in capitalism because it no longer works to the benefit of the average person. Since people are less willing to believe in an obvious failure (Occupy Wall Street, downward economic mobility, unraveling social safety net), the elite move to maintain control by power rather than consent. Thus, the elite consolidate power through a populist political regime that will suppress dissent, punish free-speech, and follow a Goebbels-type program of lies, newspeak, and "Brown Shirt" enforcement. Right-wing political orders around the world are currently moving in this direction. The Bill of Rights written into the American Constitution may become a casualty of power and falsehood. Those who have taken an oath to support and defend the Constitution (which includes Articles 14, 15, 19, and 26 guaranteeing citizen voting rights), are currently seeking to disenfranchise large segments of the poor and grant our president king-like powers.

A Personal Word

For 90% of my life, including 20 years as an ordained minister in local church congregations, I was a supporting member of evangelical Christianity. These were mostly happy years full of good times and close relationships with some of the best people one could hope to know. In many ways, I miss the relational aspects of those times.

Therefore, reader, try your best to understand that in my criticism of organized religion, especially the more fundamentalist varieties, I am making a distinction between institutional figure heads, and the people who sincerely do their best, day in and day out, to live in a manner reflective of their faith. I hold my friends in the highest regard.

As near as I can recall, my break with conservative Christianity began with the administration of George W. Bush and Dick Cheney. It had two parts: First, there was the stolen election in Florida. Although Gore had the popular vote, Bush (whose brother happened to be Florida's governor) narrowly scraped by on hanging chads. Secondly, the invasion of Iraq was built on lies and manipulation to support an illegal war against Islamics. If the basis of the war was to strike back at the main supporter of terrorism, we would have invaded Saudi Arabia, no question. But the Bush family and the ruling elite of Saudi Arabia were linked hand in glove.

Then there was the rise of the Tea Partiers, and the fossil fuel industry's efforts to build the case against global warming and the rise of sustainable energy. In each case, conservative politicians who were put in office by big money, enlisted evangelical leadership in their quest to gain and hold political power. To do this, they accented the hot-button issues of abortion, terrorism, homosexuality, the equality of women, the threat of a one-world government under the United Nations, and liberal education to define an "enemy" that would, if not checked, undermine our Judeo-Christian roots and traditional family values. They manipulated good, caring people through a campaign of fear, accusation, and doubt in order to get the vote.

Once it became clear to me that many of the leaders of national evangelical organizations were more interested in gaining proximity to politicians who, by their lives, are the picture of anti-Christian morality, I was convinced that the leadership level of Evangelicalism was a con.

During this same period, I returned to reading serious science. Since my undergrad degree is in biology, this was like returning home after a long absence. It became abundantly clear that global warming and resource depletion are an existential threat to life on Earth. It also became clear that monied interests were doing everything in their power to lie, subvert, and jam this message. And our elected officials do little to nothing to stop it. Moreover, they provide protection and cover by enacting laws favorable to industry, while harming the very people who put them into office.

In spite of these things, I am still a "religious" person in the sense that Bohr might have defined "religious." In one conversation with Heisenberg, he noted that the language of religion is different—images, myths, parables, and paradoxes—from the language of empiric science.[14] To literalize the language of religion robs it of a far deeper message, just as literalizing a dream strips away deep and subconscious truth, framed in fantasy vision, which is necessary for personal growth and progress.

So, yes, I still go to church, but not of the conservative/evangelical bent. I attend church now for two reasons: First, the language of religion carries a deep truth that helps me understand and navigate life. Some of my best thinking occurs in the context of a worship service where my mind drifts along on the notes of the organ and text of an ancient Biblical passage. I get insights of a higher order sitting there, bathed in sunlight shining through stained glass and in words thousands of years old, spoken to people like me who were trying to make sense out of life. Secondly, I go to this particular church because of the quality of people. They are more concerned with living the message of Jesus than in inspecting others for doctrinal purity. All week long, a significant slice of the congregation is involved with feeding the hungry, running a resale shop where the poor can find decent clothing at a price they can afford, providing aid and shelter for the homeless, and spreading the message of Earth care visible by the massive array of solar panels on the church roof and a campaign to inform the public about the catastrophe of global warming.

I have little confidence that institutional religion as currently organized, however, can offer a template to guide mankind through a post-dystopian world of global warming, economic crashes, and massive human loss. Looking backward from this position, the church will appear largely impotent because it was not a prophetic force to censure power and privilege. But out of these ashes something new will arise—perhaps a decentralized blend of Quakerism, scientific monism, a demythologized Jesus, Taoism, and Naturalism. The dualistic, patriarchal religions of the book will be recalled as a handmaiden of division and destruction. A new spirituality will attempt a new *re-ligio* of man to man and man to Earth. Religion, because it is part of our very fabric, will never pass away. But it will be molded into a new model—one built on an immanence that binds together rather than divides asunder. Kuhn's paradigm shift will rise from a new religious foundation—Monistic, Cybernetic, Scientific, Numinous, Inclusive, Natural, and Poetic. In short, the new paradigm will be one that has been with us since the beginning, right before our eyes, and implicated within every living system.

A **Scientific-Christian-Naturalism** works best for me.

12. Belonging

"The eye cannot say to the hand, 'I don't need you!' And the head cannot say to the feet, 'I don't need you!' On the contrary, those parts of the body that seem weaker are indispensable…so that there should be no division in the body, but that its parts should have equal concern for each other. If one part suffers, every part suffers with it; if one part is honored, every part rejoices with it."

— St. Paul, I Corinthians 12

"This fixation that I have described as an unfeeling relation of the human to the natural world is healed in its deepest roots as soon as we perceive that the entire universe is composed of subjects to be communed with, not primarily of objects to be exploited."

— Fr. Thomas Berry[1]

Another Word for Love

PSYCHOLOGISTS HAVE LONG DEFINED LOVE as the collapse of ego boundaries, such that two people in love merge into a new and unified entity. Love has everything to do with feelings of union, of interdependence, of the resolution of incompleteness, perhaps sensed more deeply at the subconscious level, but known nonetheless. One of the most basic observations regarding the universe is the law of attraction. Everything has a basic incompleteness about it and strives to fulfill that incompleteness. Mass attracts mass. Positive and negative charges attract. Molecules, by their polarity, attract one another. Unified connection is the organizing principle behind all systems.

This yearning of everything for completeness is the need to find fulfillment through belonging. Therefore, if the hallmark of love is a sense of completeness, then belonging and love are coterminous. The universe is one grand logical informational pattern, composed of a multitude of smaller logical patterns. To understand that we are a vital part of that greater pattern is to live in the context of love—it is this sense of belonging.

Let me quote psychologist James Hilman again, "There is only one core issue for all psychology. Where is the 'me'? Where does the 'me' begin? Where does the 'me' stop? Where does the 'other' begin?"[2] The healthy psyche is one that is integrated—connected—with the broader world about. Those in a loving relationship consider things affecting the beloved as affecting themselves. Greater connections produce a greater self, with the goal of becoming one with all. Ultimate loss of separate ego and merger with the broader universe has traditionally been the goal of all religions. Hippocrates recognized this 2,500 years ago, "The greater part of the soul lies outside the body" (*On Airs, Waters, and Places*).

Therefore, it is axiomatic that the purpose of religion is to reunify, to rejoin, to re-incorporate, to reconnect the individual with the greater whole. By logic, then, those who are connected to the greater universe must, of necessity, be *in-corporal* with all of its pieces and parts so that there is no estrangement, no sense of "other." Although the universe is composed of a multitude of forms, we need not be of the same form to identify with all of the others because, at base, we are all composed of the same stuff which had the same origin and functions according to the same basic laws of physics. If the nature of God is love, then the working of God could only issue in the incorporation of all pieces into one grand continuum of belonging. Truth and love—the essence of God—are connective. By inductive reasoning, the works of God become indicators of the nature of God. People have understood this for thousands of years

We were visiting Vancouver Island for the first time. After a breakfast of waffles and strawberries, my wife and I set out to see the country. Driving north from Victoria, we stopped at Goldstream Provincial Park, where I parked the car beside two western red cedars with massive girths that had developed (according to the sign) over a period of 500 years. Crossing a footbridge, I made my way along an icy path beside the stream (Joyce was cold and stayed in the car). There was an odd odor in the air, something with which I was unfamiliar. After a while, I connected the picked over salmon skeletons littering the stream bank together with the strange odor. The air, bearing the smell of fish oil, was reminding me of last fall's salmon run. Shortly, I came across a sign posted on the banks of the stream stating that fishing was permitted only for native peoples.

Driving northward again, we came to the town of Duncan, BC. Duncan is a lovely little town, quiet and unassuming, welcoming and warm. Along the main street, multiple totem poles stand as testaments to the area's native culture. In the Chippewa language, totem refers to a kinship group and is derived from an animist interpretation of the world that believes animals, plants, and places each possess a distinct spiritual essence, and that all of these essences are interrelated and co-dependent. Therefore, the totem pole is a visual statement affirming the interrelationship—kinship group—of all living things. Native peoples were Darwinian centuries before Darwin set foot on the Galapagos Islands. These totems are a simple statement in the carved images of fish, orca, eagle, and bear, saying, "This is our family." Indigenous people and the salmon taken from Goldstream are different forms of the same essence.

It's easy for us, by some faux intellectual condescension, to lump totemism with the childish beliefs of illiterate people. The Western mind does much the same with the idea of reincarnation which is central to most Eastern religions. The Greek word for reincarnation is, *metempsychosis* = "to put a soul into again." The first use of the word is attributed to Pythagoras. Reincarnation is literally, "entering the flesh again." Although the doctrines are distinctly different, the Christian concept of "born again" has many of the same tones. There is no way to prove or disprove this concept, as the Dali Lama reminded Carl Sagan when asked if he would disavow his religion if science could prove the doctrine a fiction. Nevertheless, the list of intellectuals who found reincarnation an attractive concept includes Goethe, Tolstoy, Voltaire, Jung, Wordsworth, Kipling, and Wiener.

Although the specific doctrine of reincarnation may seem like "wavy-gravy" to most, there remains some intuitive sense in nearly all peoples that we are not a mere flash in the pan, but that our life essence continues beyond temporal experience. Norbert Wiener, whose gravesite is a few miles north of this cabin, believed he was a mathematician in a former life, and that previous experience and familiarity accounted for prodigious insights that allowed him to lay the foundations for cybernetics, systems theory, and information technology. One day in December of 1954 Swami Sarvagatananda, who had recently been appointed as the new Hindu chaplain at MIT, was giving a talk on reincarnation. When he had finished, Wiener walked up to him and said, "Swami, what you have said about reincarnation, I accept it. I know you are right...I knew all these things. I recalled them when my teachers examined me because I knew these things were so in my past life. I was a good mathematician, therefore, I come now to fulfill it."[3]

To belong is to merge, one entity with another. This explains why intercourse is not only enjoyable, but necessary if there is to be any advance whatsoever. This is in no way a prurient enticement, but a logical explanation of readily observable facts. Lovers are drawn to one another out of a sense of incompleteness—we see the missing pieces of ourselves in the beloved. We merge with one another during sexual intercourse. Validation of that incompleteness is demonstrated in the frenetic race of sperm to impregnate an egg. Once incomplete in themselves, sperm and egg find fulfillment by fusing as a zygote. And the zygote is neither mother nor father, but an entirely new being derived from the recombinant code of twenty-three disparate half-codes merging into a new individual never experienced before on planet Earth.

On the other hand, most pieces of that new code are quite old, some going all the way back to primitive bacteria. Most of the genetic code within our bodies arose from archaic forms and traveled in a continuum, like the baton in a relay race, through other living things that preceded us. In a very real and literal sense, we are those orcas and frogs and eagles and bears on the totems standing today on the streets of Duncan, BC, just as we are the salmon carrying code up the waters of Goldstream. We are composites of those preceding codes, sliced, diced, edited, and recombined during intercourse over eons of time to make us, us. By the most literal understanding possible, we contain multitudes. Is it any wonder that intercourse is so irresistible, why it draws us like a magnet, and how it allows us to lose ourselves, if just for a few brief moments, by merging with another? In the act of intercourse, we escape the bondage of this illusory self and cast our information into an eternal stream carrying all who will yet come, long after we have returned to stardust awaiting a new form.

We need people. Not only for the help we can give one another, but more importantly for the ideas and insights we gain by exposure to other points of view, other experiences, other knowledge. From the earliest times, we have been social creatures gathered around flickering firelight listening to and interacting with others. As we did, our knowledge and understanding increased—we learned where the fruits were ripening, the antelope migrating, the other tribes were operating, the snakes to avoid, the mushrooms not to eat, who was up and who was down, or how to make a better arrowhead.

Learning is additive and recombinatory. Apart from this, there would be no human language, no science, no medical advances. The preeminent characteristic that marked Norbert Wiener's 45-year tenure at MIT was his weekly visits, unannounced, to professors and students in widely diverse fields to find out what was going on, what were they thinking and working on, what new things had they discovered. The man's curiosity was insatiable, and his insights prescient. Wiener openly exhibited his incompleteness and satisfied it through intellectual intercourse to make of himself one of the most creative geniuses of modern times. By merging, blending, and belonging we learn, grow, create, and increase as logical informational patterns.

The primary motivation for creativity is a sensed **incompleteness** in all things. The primary method of creativity in all things is **recombinatory exchange**. This explains the magnetic attractiveness of **uniting with others.**

The opposite of love is the Greek word, *schizo* = to divide, to split, to rend, to sunder belonging. Or, as the Hebrew equivalent is translated, "to tear apart." Doctrines of division are the manifestation of evil that breaks connections.

3-D Falsehood

The myth of the *Fall of Man* describes the greatest schism of our entire history and is the source to explain the despoliation we are inflicting on planet Earth in this current period known as the "Anthropocene." While the history of the universe, post Big Bang, is one of coalescence, joining, agglomeration, and increases in both form and complexity of being, that history hit a wall of rejection called religious and philosophic **dualism**. It is the belief, now firmly embedded in Western Culture, that man is separate from and exalted over the rest of the universe by right of divine creation and endowment with a unique eternal soul.

We were taught that Earth is not our home, that it is evil and an enticement to evil, and that our rightful home is in some conjured heaven far away. We were taught that we possess an eternal soul (perhaps at the moment of conception) that is a spark of the divine and cannot live forever in this mortal body which is corrupt and destined to corruption. We are, according to the old hymn, pilgrims in a barren land, struggling with sin and temptation, until that day when we are called to "dwell in Beulah land." It is forever and always implied that sex is evil (if not the original sin) and that sex and marriage exist primarily for procreation. "Primitive savages," we were taught, are lesser humans and little more than animals. They are not our brothers. Our responsibility was one of either subduing and converting them to the faith, or liquidation in order that proper Christian society could fill this land of Manifest Destiny. We were taught that paganism, animism, and all references to a sacred Earth are evil, Satanic cults destined to wither under the sign of the cross. We were taught, by implication, that white Christian man is the agent of God on this Earth and in God's absence has been granted *legal rational authority* to make laws and **dominate** Earth in any manner he sees fit for the advancement of Christian culture and society.

The doctrine of **death** then became the powerful tool to keep people in submission through fear. If we were good, kept our heads down, supported God's ordained authorities, and did our jobs, then heaven waited to welcome our souls in blissful reverie. If we were disobedient and challenged authority, did not live in the confines of the faith, or exposed the falsehoods of religion and power, then hell and everlasting torment would be our destiny.

All of this ethically horrendous teaching was facilitated by the primary doctrine of division. This is why, at almost every juncture, we fought, killed, and destroyed any peoples or doctrines that emphasized our oneness as humans, our relationship and responsibilities to maintain the ecosystems of Earth, and the greater interrelationship with all living things. This is the real reason fundamentalist religions cannot tolerate the science of evolution. This is why the corporate media machine does everything in its power to discredit and deny the webs of causal relationship clearly laid out by the ecological sciences.

This is why industrialists cannot allow indigenous peoples to make their case on the basis of their religion. This is why every effort and millions of dollars are spent by the extractive industries each year to discredit the spawn of Rachael Carson who would take up the gauntlet of truth-telling.

The moment we ascribe validity to the connectionist views of ecology and Earth-based religion, we must alter our view of Earth as a mere storehouse of resources given primarily for the use or abuse of man. It means the cessation of pollution. It means recycling for a zero-throughput economy. It means living in a manner that will leave the Earth pure and fertile for the next 100 generations. But this will eat into corporate profits and is a great heresy for those whose god is power and money.

Herein lies the crux of the current battle for paradigm change: Either we continue business as usual on the basis of a separatistic, divisive, **dualistic framework,** with profits overflowing in the coffers of the politically powerful elite, while life and Earth gets sicker by the day, or we adopt a scientifically proven ecological and **monistic paradigm** where man becomes a responsible steward of Earth, with the defining goals of cultivation, inclusion, and sustainable balance.

This cannot be stressed too heavily: Without the wedges of division, dominionism, and death, the ruling elite will lose privilege and power that is built on a framework of religious and political polarization. All of these divisive memes are tools of power over the masses of people who are brainwashed by fear to vote for politicians and policies that will ultimately make their lives and the lives of generations to come poorer and less healthy. I will state again—even though the elite who pull the strings of government and industry have little personal use for dualistic religion, they cannot survive without it. Divisive religion is the primary wedge they use to divide and conquer. Christians, Jews, Islamics, and Buddhists use it equally as a tool for gaining power. Wendell Berry understood this.

Writing in *The Art of the Commonplace,* Berry said, "...modern Christianity generally has cut itself off from both nature and culture. It has no serious or competent interest in biology or ecology...(and) has become willy-nilly the religion of the state and the economic status quo...It has, for the most part, stood silently by while a predatory economy has ravaged the world, destroyed its natural beauty and health, divided and plundered its human communities...in its de facto alliance with Caesar, Christianity connives directly in the murder of Creation."[4]

Without these false doctrines of **division** and separation, of human superiority and **dominance** over a fallen creation, and personal **death**, this would not have been possible. From the very beginning, in the creation account of Genesis, man, Earth, and every created thing are linked at the very level of substance. All of life is a manifestation of the "dust of the Earth." We are Earth. The animals are Earth. The grapes and wheat that nourish us are Earth. And we all return to Earth to be fashioned again in some other form.

I submit that the evidence—the fruits of these false memes—is all about us in the destruction of Earth's ecosystems, the historic and multiple failures of Christian cultures that resulted in genocides, religious wars, the relegation and treatment of fellow humans to the level of animals, the corruption of religious leadership from a Vatican that is wealthy beyond imagination, the allegedly Christian nations that justified the beating, murder, and enslavement of other humans, and the Church of England that for centuries enriched itself and provided absolution to a government that subjected half of the world's indigenous peoples to slavery, condescension, and the pilfering of their natural resources. And now, in these decades of the 21st century, the bitter bill is coming due with Anthropogenic climate change destined to result in a Permian-like extinction event, political and cultural collapse, and human carnage greater than all wars of all time. These things are the direct and irrefutable results of the doctrines of **dualism**, our assumed right of **dominance** and control, and the fear of **death**. We bought a falsehood.

And these lies bequeathed to us the disenchantment that is the hallmark of our time. Like a bad dream, the Earth's collective unconscious is showing us the horror movie resulting from illogical inputs that reject the interconnection of all things. By denying that Earth is our one and only home, we have set the house on fire and now do not know where we belong. "Rootlessness begets meaninglessness, and the lack of meaning in life is a soul-sickness whose full extent and full import our age has not yet begun to comprehend." Max Weber saw this clearly. As did van Gogh, Edvard Munch, Blake, and Thoreau. Schopenhauer confessed it that day on a German park bench, when a policeman asked him who he was, saying, "I wish to God I knew."

Until and unless we reject these false dualistic assumptions that serve only to empower the elite, we will all become sacrifices to a fiction that feeds on blood and leaves us in a waste land of disenchantment and despair.

Re-enchanting the World

Again, Max Weber understood this. Few others had a command of the scope of history, religion, and society that Weber wielded with both efficiency and extraordinary insight. He saw the trends and knew where they might lead us. Just one year before the end of World War I and Germany's humiliation by the Treaty of Versailles, Weber stood at the University of Munich in November of 1917 and delivered one of the most insightful speeches of modern times. He entitled it, "Science as a Vocation." But it was really a talk about religion and the loss of meaning in life. This age of scientific empiricism and mathematical reason, he said, had led us to a place of *disenchantment*—a term Weber borrowed from Schiller. Western culture had erected a new foundation of rationalization that issued in a modernized, bureaucratic, and secular society from which the old myths of religion and numinous nature had retreated like wood-spirits before the rising sun. Previous culture was *an enchanted garden* where magic, wonder, and awe called to man like the Pipes of Pan.

But now, the rational process had desacralized the structures and processes that formerly exalted spiritual beliefs and rituals which resulted in collective identities. The process of rationalization that caused former beliefs to fade away had now left both the individual and society with neither guides nor anchorage on an uncharted sea. Western bureaucratic capitalism had placed man in an *iron cage* of goal-oriented efficiency and rational calculation, displacing both religious connection and its attendant enchantments. We had come into a *godless and prophetless time* where conscious, linear human purpose had become the legal rational authority, displacing both God and nature as the final arbiter of right. A religion that had divided God from Earth and placed him in a heaven far away left man as the final authority for faith and practice. Ironically, the religion that displaced God to the realms of heaven, also displaced itself by making human reason the final authority on Earth. This was Nietzsche's conclusion also—Christianity first killed natural religion, then it killed its own God.

Rather than re-connect us with Earth and God, however, modern dualistic religions of the 'Book' coupled with rational scientific materialism severed the bonds that once bound us to the bosom of nature. Enchantment retreated back into the Earth and vaporized in the light of common day, leaving man as a meaningless biological accident in a meaningless accidental universe.

Does there yet remain a pathway back to Tolkien's Shire, where we may sit again with Samwise, holding our cold, tired feet near the fires of home? Can the world become, once more, an enchanted garden where we live with wonder and awe, and find the belonging that lies at the heart of love? Could we again sense brotherhood in the call of coyotes traversing the shadowlands of dusk, feel our hearts drawn upward through a piney forest toward a rising moon, or float quietly on the ocean and recognize our former home? Could we smile at death's face?

The world will become re-enchanted when we become reconnected.

Weight of Evidence

When quantum physicist Niels Bohr was knighted in 1947, he chose the taijutsu (yin-yang symbol) for his coat of arms. What message was he sending? Did the nature of a quantum universe add to some weight of evidence to convince Bohr that the universe is monist?

Biographical notes indicate that Bohr conceived of the notion of *Complementarity* while on a skiing vacation in Norway in 1927. Werner Heisenberg had informed Bohr of his forthcoming paper on the *Uncertainty Principle*. While skiing, Bohr came up with the idea that objects have pairs of complementary properties which are different aspects of the same entity, are not causally related, and cannot both be measured at the same time. He was thinking primarily of the dual nature of a photon—the smallest quantum of light—as being both wave and particle at one and the same time. Each is a property of the same entity, yet acausal in relation to each other. He would later go on to describe other complementary pairs: Position/Momentum, Energy/ Duration, Spin of Different Axes, and Entanglement/Coherence. Each member of a pair is a true description of a phenomenon that may appear as different as night and day. How can a photon be a wave and a particle simultaneously? How can this same photon have no mass, yet impact an electron with enough force to increase its energy?

The taijutsu was the closest visual approximation of what Bohr was seeing at the sub-atomic level. It is one complete unit composed of two equal portions which initially appear as polar opposites, yet function as correlates of one another. Black complements white. Their Fibonacci curves demonstrate perfect aesthetic balance, while each represents a partial description of the whole. Removal of one aspect makes the whole meaningless.

Man has long recognized that he is tiptoeing around the edge of a vast epistemological wilderness. We are reasonably confident that our description of physical reality is accurate, based on the laws of physics which are repeatedly verified by experiment. But there is another realm of life and the universe that we sense is true, that causally affects our lives, that unifies us into relational tribes, that is unseen but more real than the real, but the workings and nature of which we haven't the slightest idea. There is a shadowland between what we know and what we sense—an unknown area that Werner Heisenberg called the *Schnitt*. Theoretical biologist Howard Pattee describes it as the *epistemic cut*.[5] This *Schnitt* is so obvious, so fundamental, yet so elusive that we largely give up thinking about it. It's the darkness of *Newton's Sleep* (Blake).

For Bohr, Heisenberg, et al., it was the gap between quantum and classical physics. For biologists it is the zone between life and non-life, DNA and enzymes. For the cognitive scientist it is the unknown between objective neurons and subjective consciousness, brain and mind. For the cosmologist it is the vast question mark between baryonic and dark matter. For the theoretical physicist it is the no man's land of complementarity and entanglement. For the psychologist it is the concept of synchronicity. For the religionist it is the chasm between body and spirit, mortal and immortal. The motive force of the universe is the striving of every particular to find its meaning in balanced connections.

These are cutting edge topics for the 21st century. While we have yet little understanding of our *Schnitts*, we are making progress. And the rate of progress is accelerating and will accelerate even more with breakthroughs in genetic algorithms, quantum computing, and the science of complex systems. The understanding of emergence from recombinant logical informational patterns will be key. Understanding is a process of seeing and tracing connections—mapping the linkages of things, processes, and events one to another in the broadest possible tapestry. Since information is causal, we need to map where, how, and why information travels and influences various parts of the whole to issue in novel emergent forms and qualities, feelings and meanings.

In a previous chapter I mentioned David Quammen's wonderful book *The Tangled Tree*. Quammen's thesis revolves around the new science of molecular phylogenetics and the fresh light it sheds on the process of evolution by horizontal gene transfer (HGT). Previously, archeological remnants provided the firm ground for tracing evolutionary descent backward to determine common ancestry and origin. The late Dr. Carl Woese and his colleagues at the University of Illinois successfully advanced the concept of HGT by examining ribosomal-RNA to determine similarities across diverse species and map connections. By this method they were able to show that our mitochondria are direct descendants of a previous infection, far back in our ancestral history, by a gall-producing bacterium.

The genes that code for most of the proteins in all animal species are reasonably identical. In fact, 50% of our genes are common to a banana plant. Similar genes code for similar proteins and processes across all forms of life. By a strange twist, we do carry (relative) immortality within every cell of our bodies, going all the way back to bacteria. Most of the differences between living things can be attributed to different combinations of the same units, chemically altered expressions of the same genes, and differing transcription based on epigenetic factors.

Stepping a level higher, Gregory Bateson's observation of *the Pattern That Connects* comes into play. Structurally, all plants are built of a similar platform—xylem, phloem, cambium, cellulose, lignin, and auxins occur in different arrangements and volumes across the entire spectrum. The differences between a piece of grass and a towering redwood tree seem obvious based on appearances. But if one looks 'under the hood,' she will find that similarities far outweigh differences. Or, glance at the venation on the back of your hand and that of the maple leaf on this book's cover, or your blood vessels and the xylem and phloem of a buttercup, or the molecules of chlorophyll and hemoglobin. They are all strangely similar. Going higher yet, compare your arm with the leg of a frog, horse, dog, cat, flipper of an orca, or monkey and note the likeness. Look at the eyespot of a single celled euglena, the compound eye of a bee, and your eye. Similar genes are at work across the spectrum. Without any of the benefits of modern genetics, Darwin observed these irrefutable similarities and determined that evolution from common ancestry was a no-brainer.

In-patterning is the linkage of connection, one thing to all and all to one. Participation in pattern is the mark of belonging. Belonging is love demonstrated and the doorway to enchantment.

Going to the next higher level, we are in-patterned members of a global ecosystem. As much as we might want to believe that we can stand alone as rugged individuals in a hostile world, we are utterly dependent on this good Earth and its systems for life and continuance. A drop of water you just drank rode on a cloud from Borneo. The next breath of air you breathe will contain atoms of argon that once went through Xerxes, Alexander the Great, a brontosaurus, an orca, a bonobo, and that politician you want out of office. Last night's cod came from the Atlantic, the lettuce from the Central Valley, the blueberries from Chile, and the wine from Australia. The water I will drink at lunch comes from our drilled well, after seeping through underground veins from the base of an extinct volcano 100 million years old. This morning's grapefruit supplied energy that originated 93,000,000 miles away. Infinitely connected.

Now, let's cycle back to the subatomic world—the realm of quirky quarks, quantum wave functions, unseen entanglements, and information riding on massless photons. Let's enter the strange world of fields that distort time and matter, bend light, metamorphose energy to information and matter to energy, eat light, bore wormholes, create information by compressing matter, and form an unseen matrix that makes a monoverse out of a multiverse. In this unseen realm, a *Schnitt* stands on every street corner.

Therefore, just as there is complementarity and entanglement in the physical world, there are parallel processes in worlds of brain and mind and mental health. The weight of fact and observation leads me to conclude that the health and truthfulness of our environments affects the life and health of our minds, even though there is no clear causal connection. If the mental system of Lake Okeechobee is driven insane by fertilizer runoff from sugar cane plantations, can the people who live daily on the shores of that lake exhibit positive mental health? Can people who are told from infancy that Earth is not their home, that it is a temporal place of suffering and loss, be expected to value the health of ecosystems? Or live happily bonded to a geography of hope?

Cognitive functions point to our oneness with life and deep time. The way birds, fish, ants, bees, and bacteria operate by quorum sensing, decentralized decision making, and feedback mechanisms is nearly identical to the Bayesian processes used by our neurons to determine probabilities and take action.

The weight of evidence is clear—we are part and parcel with a grand tapestry that is life, time, and the universe. We are Earth. We are each other. We are a continuum. We belong.

Creation's Day

Sit with me now, in front of this cabin, to witness the birth of a new day ...

Night is nearly gone yet darkness rules, anchored to the heavens by silver nails. Rumors of struggle between darkness and light, born on a hint of pink light, are whispering softly just over the eastern horizon. Light, so faint, so similar to distant music barely sensed, rides on the hushed breeze gossiping among the pines. The Earth is an empty vessel, contained between charcoaled etchings of mountains undulating against infinity—formless, void, silent amid the electricity of a new day preparing for birth.

Light advances like the eye of Earth slowly opening. Monochrome silhouettes of trees and rocks rise from the land. The expanse of marsh opens, a void blanketed to the treetops in a gray and amorphous mist. Stillness is the freight of this empty vessel, its highest value contained in a nothingness from which being may arise. The air carries a scent of fertility, of wet leaves and ferns, sedges and methane bubbling from mud. Birds begin their morning chatter, thrush and sparrows seeding the air with songs that will transmute into copies of themselves.

Gray gives way to a drab gold just above the horizon. Slowly, the gold deepens then saturates into tangerine and orange. And then, as with the blasts of trumpets and coronets, the rim of the sun breaks the eastern sky with golden arrows shot from the arc of its fusion furnace across 93,000,000 miles of space. Spraying across the land, these photons of light collide with matter, infusing systems of Earth with energy from the sun. Rapidly vibrating water molecules of mist begin to warm and rise upward, slowly billowing and rolling under into waves that sink and rise again, like great ghosts of gray horses shaking sleep from their bodies. Once a formless void, some alchemic force has turned vapor into a rolling solar furnace, golden and lit from within.

Is this the prophesied baptism inspirited with fire and wind? Fire-infused breath moves across the land, a generative ether hovering above the sedges and grasses, dropping seeds of life, animating lifeless forms to motion—birds swoop in the sky, mud metamorphoses into croaking frogs. Others have fins and webbed feet. Some are feathered and flying, and a few have fur and antlers. And together, with the human eyes that watch, these fins and feathers and hooves, water, sky, trees, and mountains become a living and self-sustaining system, perfect in balance and function. Generated and moved by energy from above, this now numinous cosmos becomes a self-organized, self-regulated, balanced, and holistic being maintaining itself through the constant communication of circular feedback loops.

As the day passes, one may come to stand and gaze with awe over this singularly integrated diversity. Swallows dart in ambiguous arcs for newly hatched insects; a heron bends over the water for a fish. A mink crawls out on a rock and shakes itself. An otter carves a wake across the pond. In the evening, deer emerge from the shadowlands at the forest's edge and drink.

A man may sense an intimation of sacredness in the scene before him. Everything functions together as the most intricate symphony the universe has ever known. Each piece, without drawing any particular attention, functions as an integral part of the whole to make an ennobling song that transcends time and place. The essence is one of harmony, balance, health, goodness. It is perfected wholeness eliciting intimations of the sacred. Unsolicited, the words may rise, "Oh, it is so beautiful. It is my home. I am a part. I belong."

13. Current Crises and The New Story

"…schism in the soul, schism in the body social, will not be resolved by any scheme of return to the good old days or by programs guaranteed to render an ideal projected future … Only birth can conquer death—the birth, not of the old thing again, but of something new"

— Joseph Campbell[1]

"It's all a question of story. We are in trouble just now because we do not have a good story. We are in between stories. The old story, the account of how the world came to be and how we fit into it, is no longer effective."

— Fr. Thomas Berry[2]

"We shall not cease from exploration, and the end of all our exploring will be to arrive where we started and know the place for the first time."

— T. S. Eliot[3]

Ozymandias

I met a traveler from an antique land,
Who said —"Two vast and trunkless legs of stone
Stand in the desert. . . . Near them, on the sand,
Half sunk a shattered visage lies, whose frown,

And wrinkled lip, and sneer of cold command,

Tell that its sculptor well those passions read

Which yet survive, stamped on these lifeless things,

The hand that mocked them, and the heart that fed;

And on the pedestal, these words appear:

My name is Ozymandias, King of Kings;

Look on my works, ye Mighty, and despair!

Nothing besides remains. Round the decay

Of that colossal Wreck, boundless and bare

The lone and level sands stretch far away.

— Percy Bysshe Shelley

Paradigmatic Dead End

Well, here we are near the end and, in a tip of the hat to Eliot, find ourselves at the place where we started with Thomas Kuhn's concept of paradigm shifts. The old paradigm is finished. It's not so much that it died (ideas cannot die a natural death), but that its foundations have been eroded beyond repair and it is no longer effective. It is crumbling under its own weight. We can live in it no longer, for to do so would risk both our sanity and the life of the natural world. The paradigmatic anomalies are now so obvious and beyond denial, the destruction so vast and due to continue for hundreds of years, that to continue defending them would make us like unto madmen flailing in the desert sand to defend the crumbling structure of "Ozymandias." This is the nature of our present crisis.

"Ozymandias" always was a work of fiction, and those who

commissioned it knew that. But fictions, whether they be stories or statues, are necessary for social cohesion and for the gatekeepers of those stories to galvanize power over the masses. People of all time have been little different from that rag-tag group of escaped slaves at the base of Mt. Sinai. They had fled from the powers that erected "Ozymandias" in the first place only to fashion their own fiction as a Golden Calf in the backside of a desert wilderness as a replacement.

Most men pine for freedom but cannot handle it once it comes. Too insecure to make our own decisions, we seek some power beyond ourselves to give our lives validity. And since we know that other men are as we are—mortal—we erect idols to unseen gods. And then we go about, one way or another, abdicating and willingly transferring the right of control to a select elite who act as intermediaries between us and the fantasy gods we have created.

Aristotle compared idols to illusory reflections in water. Plato contrasted the idol to the real as falsehood is to truth. The Jews stood apart from their pagan neighbors, rejecting idols and images as illusions that fools have put in the place of the true and unseen God.[4] Perhaps Ovid (43BC-17AD) picked up on Aristotle's definition of an idol and conflated it with the self-absorption and insecurity of youth to paint his picture of Narcissus. Seeing his image reflected in water, Narcissus became fixated on it, so much that he was impotent to proceed with a mature and productive life. (Is not all idolatry a form of self-worship?) A flower given his name grew from the ground where he melted back into the Earth.

Idolatry (the worship of illusion), therefore, is a mark of self-absorbed immaturity. Fearful of facing the difficult choices, most of us would rather yield that power of choice and control to a fiction that will decide for us. Even if it requires the loss of freedoms and slaughter of competitors, we will betray truth and compromise our humanity to achieve material dominance.

At some point however, a thinking minority realizes that "Ozymandias" has no more power than any other artifice. The Kuhnsian anomalies begin to pile up. A few people begin to turn away, looking elsewhere for truth and guidance. Gradually, "Ozymandias" loses an arm, an eye, a leg, and falls face down into the desert sand.

Even though they know this god is a lifeless illusion, his promoters will spare no effort to continue this deception because it has been very profitable for a long period of time. Fortunes have been made using old "Ozymandias" as a cudgel to instill fear and guilt—tools to keep people meek and subservient to privilege—planting, weeding, hoeing, threshing, and processing the grain that feeds these bourgeois middlemen and free-riders. He is the god above all gods and the source of a people's power and position. To offend "Ozymandias," to accept that he may be a mere fiction, is to invite divine wrath, in the form of conquest by the Philistines or Babylonians or the Communists, Gays, Feminists, Evolutionists, and a United Nations one world government. These threats are mere projections of political elite motivations to control, manipulate, and gain.

The casual reader sees the clear references behind my Ozymandias rant. The military / industrial / fossil fuel / tobacco / health insurance / chambers of commerce / NRA / professional religion industries have spent decades and billions of dollars in a massive effort to subvert the truth in each of their areas of profit. And the fear is one and the same for all: If their paradigm is false, the money stops flowing into their coffers.

There is great concern over the future effects of artificial intelligence (AI). I believe those concerns are well placed. We will proceed with AI, however, for it is impossible to stop innovation. We cannot shelve AI research and development any more than we can stop the tides of the oceans. But we can erect flood gates and safety protocols. The fear is that AI (beside which we would have the brain of a mouse) will ultimately take over, sidelining humans in the decision-making process so that we would no longer have control over our lives.

But we have already done that for thousands of years by abdicating control to a political/religious system built on a fantasy. We have not been in control of our lives since we lived as hunter-gatherers 10,000 years ago. And what was the fruit of this massive realignment? Multitudes traded a life of freedom and self-determination for a life of drudgery, ill health, loss of enchantment, war, and servitude. And the problem was not AI, but another type of intelligence.

We have been on a march for 10,000 years whose end we are now experiencing in increasingly depleted ecosystems, dying oceans, global warming, and a sixth great extinction event. The force that brought us to this point was not artificial intelligence, but False Intelligence (FI). False Intelligence is our hubris thinking that we know better than the wisdom of Nature that fashioned us in the first place.

Under the "business as usual" paradigm there is no sustainable path forward. To continue with the "Old Story" is to choose suicide. The anomalies are too great. The old story has come to a dead end.

The Gods Must Die

The history of cultures is the history of stories. If one knows the story, one knows the culture. The stories of Western Cultures are mostly linear. They have good gods and bad gods engaged in a running war that never resolves on this Earth, in any person's lifetime, but is scheduled for denouement in some linear future life in some heaven or millennial period of peace when the good god finally destroys or imprisons the bad god. Although the good god is omnipotent in each of these stories, he permits the bad god to run free on Earth, destroying lives, sowing wars, violence, plagues, and massive inequities which people are counseled to endure because it is the good god's will and because he (Western gods are always male) will make it up to us in the sweet by and by if we remain faithful and only believe. The reasons for this are never fully explained, largely because they are so nonsensical. The standard bromide is that we are being tested to determine the strength of our faith and trust. Upon this basis we will be admitted to heaven or cast into hell. Or suffering builds character. Or the good god's will is a (convenient) mystery that will be disclosed in heaven if we only believe.

One the other hand, Earth-based religions put their point of reference in the only place their people know—Earth. Native cultures understand that man is a derivative of Earth and therefore a brother to all living things. This is made abundantly clear in the totemism of most native peoples, who were the original Darwinians and cyberneticians. Earth-based stories are built more closely on observation than on the fictions of Western stories. While both employ myth and metaphor with equal skill, Earth-based stories have an intelligible point of reference. Spider Woman is most certainly a mythic creation, but the overall gist of the story (as in Genesis) is that man is a product of Earth. The similarities indicate that both are stories of an original Earth-based religion. Ever since Ezra and his scribes invented a twisted rationalization for the humiliation of their Babylonian Captivity, and created Pharisaic, dualistic, separatistic Judaism and its derivatives, that Western Culture has been on a slash and burn campaign of division ever since.

Interestingly, the gods of Western fabrication never last. With every major paradigm shift, the old gods had to die. The old gods have to die because they are fictitious impediments to human progress which is ever and always based on a story. When the story becomes inadequate to explain new realities, it must be replaced with one that can be integrated with new understandings. Therefore, if the old story is outdated, then the gods on which it was built are outdated. But if the true God is eternal and unchanging, he cannot die. Thus, rather than an old god dying (as Nietzsche implied) it is simply consigned to the shelf containing children's stories. "The Big Bad Wolf," "Snow White," and "Puff the Magic Dragon" express the reality of a child's world. When the man matures, he leaves the old stories behind.

We are, at this moment in history, like an insect that is metamorphosing and vulnerable. We have outgrown our cocoon's case and crawled up on the branch to allow our wings to grow and dry. It is that very tender, very delicate moment in-between forms of being when the insect can go neither backward nor forward but must wait until it grows appendages that will take it into its next life. It's a nervous time; we feel stuck and frustrated. Our eyes turn toward the open sky above. The promise is palpable; the fears are almost catatonic.

In his film *Avatar*, James Cameron successfully contrasts the old order of dominance, extraction (of unobtainium), and depletion (destruction of the Home Tree), with the cyclical order of a place-based Na'vi culture that sanctifies nature; Eywa is a clear reference to mother Eve. Trees are central to this theme—just as sacred groves have always been holy places for native peoples. Home Tree is their anchorage to the physical order of existence, and the Tree of Souls is their access to the greater spiritual reality. Using myth to powerful effect, Cameron summarized the deep polarities tugging at Western Culture today between the old dualist extraction/depletion model and the ecocentric model of sustainability based on a re-divinized Earth. It appears that multitudes were interested in that message—the film grossed $2.78 billion, making it the highest grossing film in history to date. Can we move in that direction?

Point of Reference

Both "Ozymandias" and *Avatar* are critiques of a dominant culture. And their criticism is the same: the dominant culture, based on the dominionist narrative of consumptive extraction/depletion, is a dead end of death and destruction. A crumbled statue and destroyed environments are testaments to falsehoods. In the larger view of history, temporality and depletion are the hallmarks of failed systems. Every culture, system of government, economic, and religious system built on corrupt logic (idols) has sown the seeds of its own destruction. Proof of this is manifest in ecosystemic damage and loss of fertility that will no longer support life. Simply stated, Western cultures have always set themselves up to fail because their memetic systems are false, therefore, unsustainable. They fail because they locate their point of reference in transcendence outside the systems of Earth, based on fictitious stories with zero empirical support. This overrides internal systemic feedback which knows only truth and sets the system up for bifurcation or collapse.

On the other hand, Earth-based cultures have an immanent point of reference in feedback from Earth systems of which they are part. A coequal Earth is subject to respect and cultivation that fosters enduring fertility. This places man within, not above, the government of Earth. What is good for Earth is good for man. As a result of a deeply entrenched "land ethic,"[5] subsequent generations of life can continue existence as part of a sacred unity to which they belong (see, David Suzuki, *The Sacred Balance*, 2002).

This paradigm shift is absolutely necessary to avoid a global dystopia of our own making. **We must shift our point of reference from transcendence to immanence.** This does not represent a reversion to some barbarian/yahoo culture. It means recognizing that we are part and parcel with one grand cybernetic system that derives command and control from internal, circular feedback, and not from the heavenly realms of a sky-god. The system knows what is best for the system. If we heed the wisdom built into the system, we will make the right choices. When honest information is causal for choice, outcomes foster continuance. Ignored or denied feedback is an invitation to failure.

In *The Big Picture* (2016), physicist Sean Carroll likens the cultural shock of paradigmatic change of reference to the old Road Runner cartoons featuring his eminence, Wile E. Coyote. Inevitably, Roadrunner (being the smarter of the two) would lure unsuspecting Wile E. on a speeding race to the edge of a cliff, which Roadrunner would dodge at the last minute. Poor Wile E., however, doesn't recognize the cliff edge until he overshoots and is suspended momentarily in mid-air with his feet still running. Then, with a look of absolute horror, he plummets earthward in a heap.[6]

In the old paradigm, we do not belong, Earth is not our home, life is a struggle against evil within and without, and we are permitted to do as we wish with Earth since it is passing away and will be replaced by some transcendent new Earth that just drops down out of heaven. People who are divided on the inside will leave their world divided on the outside. Man will ultimately transform his external environment so as to reflect his internal environment. Reciprocity works in both directions and is one of the best indicators of either health or disease, truth or falsehood.

The Near Future

No person knows the future. I like the quote, variously attributed to Niels Bohr and Yogi Berra: "I never make predictions, especially about the future." For Bohr, there was just too much uncertainty baked into the quantum wave function, and for Berra the future depended on the way the ball bounced, and nobody could predict that.

The next two decades stand to produce some of the biggest changes in all of human history. That sounds fairly dramatic, but I believe it is an accurate statement based on the accelerating rate of technological change, human population growth, global warming, and the abandonment of theological dualism. Let's look at them in that order:

1. Technology
Wentworth-Douglass Hospital, May 16, 2019

It's the standard hospital waiting room—padded chairs, magazines, people fidgeting nervously as they await news, an antiseptic smell. It's now 7 AM and she's in surgery. I go to the cafeteria to get an omelet and coffee, with a second cup nursed along to kill time. Back to the waiting room. The doctor comes soon, smiling and chipper. "It went well. She is strong and should make a speedy recovery." Then he's on his way to replace another painful joint.

My wife has endured years of pain, sometimes worse than others, and finally decided to have that bum knee replaced with a stainless steel/plastic contraption that won't become arthritic. She will return in June to have the other knee replaced. With proper care and exercise, she may walk pain-free for the first time in years. It's quite a relief for someone so used to activity.

When the surgeon returns to the OR, he may do another knee, or he may replace a hip or shoulder. And he's not alone in this human refurbishing business. There may well be other surgeons in the same hospital replacing hearts, kidneys, or lungs. A team of brain surgeons may be inserting electrodes in someone's brain to alleviate chronic pain or address seizures. These types of surgeries are accelerating rapidly.

This is only the beginning. We stand now on the threshold of procedures and alterations to the human body that were once the stuff of sci-fi. And joint replacements are like playing with Legos compared with what's happening in major university research facilities as I write. The explosion of knowledge in the fields of information science and biotechnology could change life to such a degree that we may well be on the cusp of a giant leap in evolution—the evolution of non-carbon based bodies powered by consciousness held in artificial neural networks that can compute and recall 1,000 times better than the old, carbon-based dinosaurs they replaced—**Homo computronium**. If you watched the films, *Ex Machina* or *Transcendence*, watch them again, because they may be a future reality that is coming far faster than you think.

Proteins formerly non-existent on Earth are being engineered in labs using evolved bacteria as constructors. Genetic editing is almost old hat. Scientists invented a neuron and are working on artificial neural networks similar to those in lower forms of life. AI experts estimate that we will create human level intelligence by 2047. Meanwhile, computing power is rising exponentially. **What used to fit in a building, and now fits in your pocket, will fit inside a blood cell.**[7] Google's Alpha Zero, a product of Google Brain, became a chess-playing wiz by teaching itself to play in four hours. IBM, Baidu, Apple, Google, Microsoft, Alibaba—all of these corporations are heavily invested in two things: massively increased computing power, with emphasis on developing Quantum Computing which will make today's mainframes look like an abacus, and genetic algorithms with the goal of developing one Master Algorithm that can absorb all human knowledge and teach itself anything. And the rate of progress is accelerating exponentially. Forty pencils added linearly are forty, multiplied exponentially they are a trillion. Consider this mind game ...

You are in a hospital bed, old and dying, wired to a scanner. For 80 years you have lived a mostly happy, mostly healthy life, but now the old body is just worn out. Your immune system is weak. You are in the last stages of pancreatic cancer. Organs are beginning to fail. The prognosis is dire. Beside you sits an inanimate, non-carbon body identical to you at aged 25 and hooked to a computer. Within the synthetic cranium sits a new mass of artificial neural networks vastly superior by every measure to your old brain. Technicians do a million scans of your 80-year-old brain, gathering every bit of information you hold, then hit the enter key. In a few moments you stand and look quizzically at the dead body that used to be you. You walk out of the hospital feeling like a new person, and you are. It's a beautiful day in early summer so you make your way to Wolfeboro Bay and sit on a bench as the *Mt. Washington* docks. You and your friends are going on a dance-cruise ...

Science fiction? Yes. But at one time, so were airplanes, rockets, a man walking on the moon, and a computer in your pocket that is more powerful than an old mainframe.

As a result of advances in genetics and biotechnology, disease treatment will soon move from shotgun blast medication and radiation, to rifle-shot gene and immune therapies. Carried by bacteria, gene fixes will be injected into the bloodstream, find their target, and replace the faulty gene causing the negative manifestations.

It is not beyond speculation that neurosurgeons will soon be implanting modules in the brain to repair lost functions. Spinal cord injuries may become a simple matter of inserting the proper stem cells in the damaged area. The above list will prevail ... but for a short time only.

Even in its nascent stages, artificial intelligence is doing all sorts of things better, faster, and cheaper than human counterparts. Algorithms read body scans with 90% accuracy for detecting lung cancer, compared with a 50% rating for doctors.[8] A San Francisco pharmacy, using a robot, filled two million prescriptions in one year without a single error, compared with an error rate of 1.7% for human pharmacists.[9] The January 12, 2019 issue of *Business Insider* ran an article stating that by 2030—10 years away as of this writing—artificial intelligence and robotics will replace 800 million workers. Autonomous vehicles will have a huge impact on delivery and taxi services, although safety concerns will keep human drivers on for some time. However, after so many billion miles of driving with a much lower accident rate, it won't take long for profit-driven capitalism to replace human agents.

All of this will force a rethinking of the role of human governance relative to the old orthodoxy of free-market capitalism. If millions of people are suddenly no longer needed, what to do with them? They will have to eat and live somewhere but will no longer have the means to secure the necessities of life. Do we euthanize them? Let them fend for themselves in the streets? Marshall the National Guard to keep them from looting for food? Suffer a revolution? Or do we initiate some form of Guaranteed Basic Income, partnered with higher taxes on the rich? Can you see the current Senate voting for that?

There will also be massive changes based purely on old model inefficiencies, especially in energy and health care. Now that photovoltaic panels are competing on par or lower than coal-fired and some natural gas fired generators, it is only a matter of time before the old energy paradigm is left in the dust. Couple this with improvements in battery storage technology and the handwriting is on the wall for transportation powered by internal combustion engines. Perhaps this will not lead so much to overall job losses as it will to financial loss. Those hundreds of years of coal deposits and a century of oil deposits yet to exploit will become stranded assets because they are not needed and are a major source of climate warming. Fossilized stock funds, retirement accounts, and their massive supply chains will all be hard hit.

Regarding health care, people will finally come to realize that the current system is loaded with inefficiencies that lift costs by up to 50%. Medicare can process a claim for $2, whereas private health insurers' costs are around $15/claim. Once people see private insurance as the tax it is, and that it costs twice as much as a Single Payer System, the political momentum will shift, and millions of jobs will be eliminated because they add little to no value to the system.

Adding it all up, a serious conflict is in the making. On the one hand, the global population is increasing. On the other hand, fewer and fewer people will be needed for current levels of production. How will governments handle this conflict? Can we avoid a global revolution?

2. Human Population Levels

In 1944, the 19 coast guardsmen occupying St. Matthew's Island in the Bering Sea decided they'd like to have fresh meat. So, they imported 29 reindeer and turned them loose to feed and reproduce on the 128 square mile island. It was Nirvana for the reindeer, with abundant food supply and no natural predators. Thus, they ate, reproduced, ate, and reproduced. Population numbers increased geometrically but the food supply did not. The few reindeer harvested by the guardsmen barely made a dent in the population. The base closed in 1946 but the reindeer remained, doing what reindeer do. By 1957 their numbers had increased to 1,350, and they were eating themselves out of house and home. In 1966 only 42 remained—weak, emaciated skeletons rattling across the land. Finally, in 1980, having scoured-up any remaining vestiges of nutritious vegetation, the last reindeer died.[10]

The Rev. Thomas Malthus (1766-1834) would have said, "I told you so." Growing up in one of England's more literate families,

Malthus took honors at Cambridge and was ordained an Anglican cleric. His keen mathematical aptitude, interest in science, and writing skills earned a Royal Fellowship. In his 1798 *Essay on the Principle of Population*, he argued that populations tend to increase geometrically and overshoot their means of subsistence. In the absence of natural controls, famine and disease cull population levels until they return to sustainable balance with resources. The four-year lemming population cycle is a perfect illustration.

In the 1960s, Stanford professors Paul and Ann Ehrlich applied Malthus' principles to human numbers in their book *Population Bomb*, in which they warned of imminent famines and population crashes. Yet, it did not turn out that way, at least on a large scale. At the same time, Norman Borlaug pioneered the "Green Revolution" by developing high-yield, disease resistant varieties of wheat cultivated under the best agronomic processes. Crop yields skyrocketed enough to feed another billion humans and stave off any crash predicted by the Ehrlichs. The naysayers laughed at them, relying on the assumption that science will always find a way, in combinations of genetic engineering, better field practices, more efficient distribution, and more vegetarian diets. And to a degree, they are right. The prediction business is a minefield.

On balance, however, agricultural yields are showing signs of decline. GMO crops often underperform traditional seeds and methods. Herbicides, while temporarily suppressing weed competition, have facilitated the rise of "superweeds" that require even stronger herbicides, many of which are extremely toxic to animal life. Globally, topsoil loss is occurring 10X faster than replacement. Warming climate is decreasing agricultural production in tropical/subtropical zones, resulting in masses of climate change refugees. This has become a political bomb in Europe and America and the force behind rising nationalist politics, laying the lie to those who say population growth is not a concern. Food scarcity and politics has recently starved 86,000 children in Yemen, with multiples of uncounted others globally. If population increases are no problem, why would nations close their borders? Let's look at some hard numbers.

A random internet search yielded the following results for human population numbers at certain points in history: Ten thousand years ago, at the start of the Agricultural Revolution, there were about five million people on the entire Earth. By AD 1000 those numbers had skyrocketed (as a result of more available calories due to farming) to 350 million. The graph continues to rise, wavering a little around the time of the Black Death, but stands around 750 million when Europeans began colonizing what is now the United States. The Industrial Revolution gave a huge boost, sending populations up to 1.6 billion in 1900. And then things just explode during the next 120 years to the present where the global population is at 7.4 billion mouths to feed. In just over a century, the global population has grown explosively by nearly 500%. The Ehrlichs were correct to call it a *population bomb*. And that detonation, more than any other single factor, is the cause of the critical demise in all of Earth's life support systems.

It is virtually impossible for the entire 7.4 billion people on Earth to enjoy a lifestyle common in the developed nations of the West. In their landmark book *The Limits to Growth: The 30-Year Update*,[11] Meadows, Randers, and Meadows employ systems analysis methods to estimate that on a broad scale, humanity began to use resources that exceed Earth's regenerative capacities in the year 1978. We are currently using up around 1.5 Earths. I did a search to see what brave souls might offer some concrete number of the people who can live sustainably on Earth. The numbers were all over the place, from 2-40 billion and more, with little documentation. It's almost like talking about limiting human population numbers is equivalent to grabbing the third rail—sure political death.

It seems that current trajectories offer a better picture than Ivory Tower speculations. First off, agricultural production drops precipitously as temperatures increase. For instance, demand for wheat is due to increase 60% by 2100. Meanwhile, as temperatures rise rapidly, production drops by 6%/1° C. But that decline is not linear. Above 103°F, photosynthesis stops because the enzymes that keep it going deform and cannot function. Tropical and subtropical areas are due to become food deserts as temperatures continue to rise. Attendant water shortages will only compound this as millions of square miles turn into de facto deserts. And this is just the beginning. It is happening today in Africa, the Middle East, and is moving to China and India.

Across the board, yields for nearly every crop grown in California's Central Valley will decline due to higher temperatures. Some crops will no longer be economically viable. Snowpack in the Sierras, which provides needed water for irrigation, is projected to decline 65% by 2100. Meanwhile, if estimates given in an earlier chapter about topsoil degradation and loss prove true, much of our fertile topsoil will be gone by 2100.

The upshot is quite simple: Populations will be increasing while food production is decreasing. We might have 80 years before average global temperatures rise at least another 5°F, and rainfall declines in many current agriculturally productive areas.

We can reduce population the **hard way**—war, starvation, disease, and genocide. Obviously, there is no way to square these methods with the doctrine of any accepted religion. Moreover, they are betrayals of the very doctrine and spirit of all religions. They are Satanic. Yet, culturally accepted religions are pushing the world in that direction by denying both the problem and valid scientific consensus. Religion could be a prime source of human suffering.

Or, we can choose the **easy way**—by engineering a soft landing through public policy that encourages fewer births, easily accessible birth control, and educational programs that both detail the disastrous results of overpopulation and encourage people to take responsibility for their fertility. Aid to foreign nations would be contingent on acceptance of these policies and audited proof of operational programs. To enhance the effectiveness of these programs globally, advanced nations would devote funding to universal secondary education with bonuses for achievement. The basic idea, especially for developing nations, is to have fewer children but invest more in them. As numbers decline, life becomes more valuable. We dedicate more resources to raising the children we have, ensuring both them and the Earth a better, more peaceful, more prosperous future. Besides, if artificial intelligence is going to eliminate millions of jobs, we will need fewer people, and they will have better lives.

3. Global Warming

Although this topic has received extensive coverage in other sections, we have yet a few things to consider. Global warming will be the main driver of societal and ecosystemic bifurcation going forward. Few problems—economic, demographic, political, military, agricultural,

cultural, or religious—can be considered in full without addressing the massive impacts of a climate that may soon enter a phase of runaway warming. Once temperature rise achieves runaway, not one thing we could do will stop or reverse the trend. Even if we stopped burning fossil fuels today, consensus among scientists is that Earth's atmosphere would continue to warm by another degree F. This is because of the lag effect from ocean warming. Therefore, it appears that weather patterns of the 2030-2040 decade will be the new normal for thousands of years, regardless of what we do. Unfortunately, we still do not know exactly where the tipping point is for the "Methane Bomb." Once we cross that tipping point and tundra permafrost begins to thaw releasing methane from the vast stores of now frozen organic matter, we are helpless to stop it. This will create a positive feedback loop that will release even more methane at a faster rate yet. The "methane bomb" will be like an avalanche racing downhill, ever faster and ever larger. Civilization is the garden at the bottom of the slope.

Increasingly violent swings in weather events, decreasing agricultural production, massive climate induced human migration, resource wars, and nationalist politics appear to be baked-in for the foreseeable future.

4. Theological Dualism

It's just a matter of time before Western man transitions from a worldview of theological dualism to a spiritual/scientific/ecological monism. As the foregoing list of attendant anomalies pile up, people will understand that these failures are a direct result of a religious/cultural dualist dominionism. With each passing day, it becomes clearer that man and Earth, life and information, feeling and causality, mind and body are one grand, interconnected, semantically linked, circularly causal unity. For the first time, astonished as though we have stumbled onto some grand philosophic/religious breakthrough, we will come to understand the way Earth and life have operated from the very beginning. In a strange twist, we will have achieved the ultimate goal of every true religion—the integration of the one with all. **We will have returned to where we began and know the place for the first time.**

Nations are now little more than fictions in this age of globalization. Multinational corporations are the real governing force. Through their political appointees, they write the laws. They run the internet. They run the banking system. Through the World Bank and in the name of progress, they shackle undeveloped and developing nations with debt in order to get access to their resources. The problems of political instability, crime, coups, and asylum-seeking refugees from Central America are directly traceable to manipulations by American corporations through the CIA.[12]

The New Story

New stories, new paradigms, new world views, new technologies, and even new people are never totally new. Every new thing is a composite of older pieces and parts, ordered in novel combinations, that have been kicking around for ages—"What has been will be again, what has been done will be done again; there is nothing new under the sun" (Ecclesiastes).

Yet, newness is everything. Although the pieces and parts of a new order may, in fact, have been around for a long time, a new ordering endows them with novel emergent properties perhaps never seen on Earth before. Silicon is ubiquitous in nature, but never has it had the power and abilities it does when patterned into semiconductor chips. Life and the world advance on the emergent qualities of these new, more complex patterns.

Let me be clear—I do not advocate dumping all of our cultural and religious stories, with their myths, parables, and fantastic visions, for a totally new faith that calls us to glibly trust science, go on a vision quest, reach out to the unknown, and hope for a better future. These stories give us a richness of insight into who we are and how we think, along with the positives and negatives that brought us to this present moment. They contain deep lessons that can serve us in the future as we go beyond our limited understandings and misinterpretations, peel away old illusions, and discover new kernels of truth. Interpreted against the backdrop of new knowledge from the fields of biology, genetics, quantum physics, informational sciences, cybernetics, cognitive science, philosophy, ecology, and linguistics, the old stories often metamorphose into winged visions. Otherwise, how could people with both feet in the information age still find meaning in the lyric tales of Homer, the Greek tragedies, Shakespeare, or the parables of Jesus? Real diamonds are multi-faceted translators of light into every conceivable hue. Yet, they are just common carbon, highly compressed. But every time we change our point of view, they yield illuminating novelty.

In no way does science negate this story. On the contrary, Darwinian evolution, physiology, archeology, chemistry, physics, and molecular phylogenetics are just fleshing out the details of a story we could formerly understand only in fantastic mythical form. Without the aids of modern science, the language of myth was our best summary and approximation of truth.

That truth has two parts—matter and causality. In terms of material substance, we, and every other life form, are just different combinations of common Earth with differing qualities and abilities ranging from the simple to the complex. Evolution and theistic creation stories both proceed from lower forms of life to higher. We share commonalities of structure, function, energy needs, foods, environmental requirements, and cyclical lifespan. In short, we are all related within the family tree. As previously suggested, take a walk-through forest and field and learn to see every living thing you encounter along the way as a prototypical iteration of you, and yourself as a manifestation of the potential they carry. We are all cousins. Remember James Hillman's question: "Where does self end, and the other begin?" We know that our long night of separation has ended, and the new day has begun when we can look at Earth and in the face of every other life form and see ourselves.

Re-divinized Is Re-enchanted

Is there yet a way out of Weber's disenchanted world of rationalization and technology that strips our humanity, leaving us in an *iron cage* surrounded by the *icy darkness of polar night*?

I believe there is, and that is the goal of this book. **My main hope is simple: A reconnected world is a re-divinized and re-enchanted world where we belong once again.** If disenchantment is the end product of removal of God from the world, then moving God back into the world will re-enchant it, because enchantment is the by-product of reconnecting with the deep and unseen causal truths and forces operative in this world. Enchantment is knowing that we are bound, by matter and spirit, to every other thing, life, person, and force on this green Earth. Enchantment is looking at a totem pole and understanding the literal truth it represents—every bit as much as the logic of Charles Darwin, Niels Bohr, Gregor Mendel, Aristotle, Watson and Crick, Spinoza, Carl Jung, Werner Heisenberg, Walt Whitman, William Blake, and Albert Einstein connects us, cell and bone, to every other life and particle across all of time and the universe. If we are to re-enchant the world, we must bring God out of exile beyond the clouds and return him to the heart of every atom, every bird that soars, every cloud that drifts across a cerulean sky, every sun that rises and every plant that reaches for it, every smile that greets us, and every tear that falls. Was this loss of connection Nietzsche's pain? Is it not the source of all madness?

I do not know. But some consider that the end of Nietzsche's jeremiad against the world of his time was a longing for a re-divinized Earth.[13] He was more than willing to rub the establishment's nose in the death of its gods, because they never were gods, but mere fabrications of our striving for power and control. Divinity is at the heart of the world, but the priest, the technician, the politician would not see it because that Truth conflicts with the idolatries of money, power, control, and entitlement.

According to Nietzsche, it was the artist who would see and point out the divine that lies at the heart of all things—people like Blake, van Gogh, and Whitman. The noble man of the future, therefore, would be the artist who would make the poetic connection between life, matter, and Immanent Deity. The world and time are not linear as some suppose, but cyclical and ever renewing. Both philosophically and religiously, we live in a world of the *Eternal Return* where deity is at the heart of everything. Nietzsche was certainly not the first to see this.

Long ago, Heraclitus observed that, "All things are one." The word "universe" means "all things turned into one." The Hindus believe in reincarnation, while native Americans believed the animals were their ancestors. The heart of every mystical experience, whether religious, near death trauma, or natural, is an identification—a oneness—with all of creation and the loss of a fear of death.

Both Nietzsche and Gregory Bateson considered themselves atheists in the traditional sense. And both arrived at the same conclusion: The artistic/poetic vision is the pathway to a reconnection with the oneness of life and Earth, and may be the nearest thing to a religious experience capable of rousing us from an insanity that is driving the ecosystemic destruction now threatening to destroy life and civilization as we know it.[14]

I submit that scientific truth is also a viable pathway to re-divinize the universe. In recent decades, scientific endeavor has gone from an isolating reductionism to an inclusive and holistic approach based on pattern, field, and connection. The logic is simple: Nothing can be fully understood in isolation. To know a thing is to understand its patterned connections to every other thing. The epidemiologist can neither understand nor make any recommendations regarding coronavirus without knowing the human immune system, the ecology of disease vectors, global travel, and the attendant agricultural economies of southeast Asia. A man who comes down with the Covid-19 in Peoria,

Illinois is indirectly linked to bats and pangolins in China.

Likewise, loss of tourism dollars in the Florida panhandle is linked to dead fish that resulted from an outbreak of "Red Tide" that was caused by fertilizers that came from corn fields in Iowa that were subsidized by public policy that was influenced by lobbyists for corporations that make ethanol who claim it makes us less dependent on politically repressive regimes in the Middle East that were once destabilized by our CIA in order that giant oil companies could make more money but caused them to distrust us, radicalizing some who commandeered commercial jets to crash into the twin towers of the World Trade Center that created a rationale for another war in the Middle East that deepened their hatred of America and led them to fund militant Islamic groups that have bled America of $5-7 trillion and may hasten our bankruptcy because we wanted to make more money. Ironically, the drive to make more money may be the precipitating source for our insolvency. Understanding connections is everything in healing.

The primary effort of science today is teasing out the linkages between all manner of seemingly unrelated processes and manifestations. At one time, people would have considered the suggestion that fertilizers in Iowa could cause economic loss along the Gulf Coast a laughable fantasy. Likewise, it would have sounded ridiculous to say that government efforts to ensure the profitability of an energy company were impoverishing the world we live in—desire for wealth leads to poverty. Once the connections are clearly understood, however, these statements become logical. But let's not be too hard on reductionist science, for it helped lay out more of the pieces of a puzzle that other science is now beginning to reassemble into a grand tapestry that is paving the way for the re-divinization of the world.

Therefore, multi and trans-disciplinary modern science, aboriginal mythology, artistic vision, and ancient Greek Logos theology are converging to bring God back to the center of every piece and part of this world's systemic goings-on. The truths of these four diverse points of view are now coming into a unified phase like the wavelength of a laser in a searing light that will vaporize illusions of division, separation, and transcendent causality.

Having traveled a long and arduous road that made its way across empires and civilizations, along a twisted path full of dead ends and rejected scientific theories, through theological volumes that were little more than speculation, and wars for a god nobody really knew, we will return where we started around a communal fire. We will have returned to our only home and to the God who is immanent in all things, the God who is Meta-Logic, who is One, and at the same time, Multitudes. A re-divinized world will be re-enchanted. We will again be home, in the only place that can be our home, the only place where we truly belong.

What Will the New Story Look Like?

How will life under the new paradigm differ from that of the early 21st century? What will our new institutions, new cities, new transportation, and energy infrastructure look like? How will we mitigate global warming, acidic seas, and compromised ecosystems? How will we bring the human population back to a sustainable level relative to global carrying capacity? Obviously, answers to these questions could fill volumes.

There is one major qualifier, however. Unless we can overcome the political gridlock that keeps us running in place, squabbling with each other, and destroying Earth in the process, nothing is going to keep the world from apocalyptic chaos. Continuing business as usual will make every one of the following suggestions futile. We must recognize that business as usual is a choice—it is the choice for global catastrophe of a magnitude mankind has never witnessed. To choose business as usual is to choose the hard landing where we wreck the airplane and kill most of the people on board. On the other hand, changing leadership from the current global kleptocratic elite to a life-centric order might possibly enable us to land more gently on the water. Although we can never go back, we can limit damage and loss, and then begin the process of rebuilding a better world on a better truth. When we finally come to believe that Earth is our only home, we will treat it as such. To love Earth is to love self and others.

New Religion: The word "religion" makes some people uncomfortable. Many see religion as narrow, constrictive, an insult to intelligence, a fantasy, the bailiwick of charlatans and impostors, the force behind innumerable wars and ethnic genocides, and a pablum for the gullible. I understand these points of view. The anti-religionists are merely calling out what they see. They are speaking the truth. Yet, to deny the religious impulse is to deny much of our shared humanity. Frans de Waal, a confessed atheist, argues that the roots of religion are readily observed in the moral and ethical rules of primate tribes. The instinct for religion has evolutionary roots. De Waal then goes on to quote Freud, "If you wish to expel religion from our European civilization you can only do it through another system of doctrines, and from the outset this would take over all the psychological characteristics of religion…"[15]

The religious impulse is such a central part of our being that we delude ourselves by thinking it will ever go away. As long as man is a social animal, he will need a moral and ethical system by which to order his society. Society will not work if everybody is playing by his own rules. Cooperation would shut down because there would no longer be a basis for reciprocity, for ethics, for justice, for mutual protection, care for the vulnerable, or business transactions.

The basis of religion is delineating and choosing the connections, attitudes, associations, and behaviors that foster mutually beneficial relationships which provide thermodynamically positive outcomes for long-term survival. Therefore, individual and societal choices that decrease greenhouse gases, stop pollution, generate equality, reduce population through declining birth rates, result in transparency, demand equal justice, reject war, demand zero-throughput economies, and seek to forge caring connections between peoples and the Earth, have high religious/spiritual value.

Thus, the job of religion going forward will be to choose and accent memes that contribute to long-term survival. Of necessity, these memes will be Earth-centric, homeostatic, redemptive, inclusive, transparent, intelligible, just, humble, value feedback, eschew falsehood, and connective.

What will the new thing resemble? What will a new understanding that values positive thermodynamic outcomes and rejects the dead weight of hypocrisy and unintelligibility look like?

It is impossible to predict the final form—if there could be a final form—of any new religious platform. But I see several possibilities for the mix in existing forms. Let me suggest a few ideas:

I believe the new age will be decentralized and distributed, while under a loose confederacy for oversight and coordination. It will not be hierarchical but will emphasize equality and the value of individual insight, much like the Quakers who believe in the inner light of each person.

The re-divinized Earth will play a central role. Earth, as a generative source, will be honored through doctrinal demands that decry waste and pollution, encourage a respectful, humble, and protective attitude to all other life forms because of our relational connections, and exalt sustainable values. Connectionism will figure into spiritual practice, with Earth respected as a sentient being capable of communicating valuable life lessons to man via natural feedback mechanisms. If the Earth is sick, if an animal population is disappearing, we will examine our attitudes and behavior to determine if we have played a causal roll. Future religion will demand remedial actions based on cause/connection mapping. Damage to Earth will be seen as religious apostasy.

A new religion will be Monist. Mankind will have rejected the unintelligible assertions of dominionism and dualism that feeds on the fear of death. We will put nothing into a river that we would not put into our mouths. We will release nothing into the atmosphere that we would not want released within our homes. God and man, nature and the universe, will finally be understood as one interconnected unity. We will have returned to our roots. We will be home and belong again within the entire family of nature.

Perhaps reflecting a personal viewpoint, I believe that Jesus will be seen in a new light, not as a transcendent deity, but as the ultimate pattern for a noble human life, worthy of honor and emulation. By a

Jeffersonian attitude, we would see the attributions of the miraculous as parables, metaphors for the deep healings we need. This life of utter humility, unselfishness, inclusiveness, and lack of concern for material gain will provide a strong antidote to a consumerism that is currently killing the Earth. Prophetic courage in speaking truth to power will mark religion's call for accountability in leadership. Would not the world be a better place if we literalized many of his teachings—turn the other cheek, care for the sick and destitute, bless your enemies, don't worry about tomorrow, reject the sword, feed the hungry, walk humbly, reject selfishness, do not fear death, and rejoice in truth?

In many ways, scientists are the unwitting handmaidens of a new religion. Every day, scientific research is revealing new connections between everything. If religion is re-connecting, then scientists are doing far more than an army of priests and preachers to empirically demonstrate that we are, in fact, connected to everything and every process in the entire universe. Science is returning us to our place in the only home we have ever known—not in some make-believe heaven far off beyond the clouds—but right here, right now, on this green and wondrous and deeply interconnected Earth. Science, once the disenchanter, is re-enchanting the world once again.

So, this is how I see it: the old myths reinterpreted for the information age, early Greek Logos theology, the human pattern of Jesus' life and teachings, aboriginal Earth-based religions, an integrationist, pattern-based science, abandonment of hierarchy, and a Monist view of the universe will be the syncretistic scaffolding for a new religious integration of Earth, man, and universe. Structure and practice will emerge from newly patterned values and information.

New religion is placed first in a list of new structures. This is because

I believe most human systems and attitudes stem from our core world view. The systems that follow are derivatives of our most important values.

New Culture: Cultural emphasis will shift, from having to being, from consuming to cultivating. The world will be a more egalitarian place where we finally put the guiding principle of the Declaration of Independence into practice—"that all men are created equal." Culture will be more integrative and less separatistic. Believing in the power of human potential, public policy will emphasize and fund all education through college. Open collaboration will promote trans-national alliances for mutual benefit. Nationalism, ethnic superiority, and racial prejudice will be seen as regressive and discouraged. Having multiple children will be seen as selfish and discouraged by public policy in favor of two or less children/couple. Unless we cut birthrate to less than two, we have no hope of avoiding an apocalypse. Free access to contraception and sexual counseling will be available to all. Subsidized childcare, Medicare for all, and a guaranteed basic income will become standard.

New Economics: The new economics will require absolute financial transparency of all elected officials, banks, financial exchanges, and corporations. Dark money havens will become illegal, under threat of economic and travel sanctions. Shell corporations will be eliminated. There will be one global tax rate. Trades on financial exchanges will be delayed 10 seconds to give all equal access. All derivatives must be reported, as well as all hedge fund and investment banking holdings and transactions. In the spirit of free-market capitalism, banks and exchanges will be required to hold all risk associated with their investments. Losses based on bad investments will first be deducted from the incomes of salaried employees and members of the boards of directors, and then from shareholders. Proven fraud will result in loss of license to engage in public global financial activity. To break-up financial dynasties, no legal heir may receive more than $1 million in total combined assets. All the rest will go into the US Treasury to fund education. Polluters will carry 100% liability for any environmental damage they cause.

New Politics: All election campaigns will be financed equally and solely by the US Treasury. No dark monies, no media asymmetries will be permitted. The Fairness Doctrine will be revived and applied to all public media. False statements about an opponent will be subject to existing libel laws. Campaigns must focus on issues and will be limited to a period of three months prior to the election. All public officials will be elected by popular vote. The Electoral College will be eliminated. All voting districts will be delineated by unbiased committees that have no beneficial interests. No former public office holder may, at any time, serve as a paid lobbyist. No person from industry may serve on the regulatory board that supervises his or her industry/business. A two-term limit applies to all elected offices.

New Transportation System: The goal will be 100% electrified transportation. There will be a national network of electric rail for both passenger and freight. Likewise, cities will adopt policies phasing out all fossil fuel vehicles by 2030. Local electric rail will tie in with and support the national network of high-speed rail. We will eliminate fossil-fueled jet travel by 2030 and develop faster sea-going ships that combine electric motors with wind power. Overseas supply chains will be taxed according to their pollution, then reduced as carbon neutral manufacturing returns to each nation.

New Energy Infrastructure: The nation's energy infrastructure will be based on renewable energy from wind, water, photovoltaic, geothermal, and wave/tide energy. In order to distribute surpluses to areas of need, the government will sponsor a massive upgrade in the national grid to accommodate long-distance transmission from areas of low population density to areas of high density. Every public/corporate rooftop and parking area will become available for the installation of photovoltaic panels connected to the main electrical grid. There will be a national net-metering policy. All subsidies to fossil and nuclear energy will end as of 2025, with those subsidies going to fund battery research and development, infrastructure upgrades, high-speed electric rail, and domestic photovoltaic based on financial need.

New Cities: Since populations are becoming more urban, we will engage in a massive program to tear down old and inefficient cities, repurposing all salvage materials for new structures, to replace them with energy efficient, human-centric cities. My model city would be pedestrian only, with at least 160" separating rows of buildings six stories or more in height. The first floor would be business/public space, with five stories of apartments ranging from 800-2000 square feet above. Pedestrian only (with al fresco dining) sidewalks would each be 30' wide (60' for two) leaving 100' between buildings and sidewalks for public gardens, water features, arboretums, play areas, and dining. Electrified handicap transportation would be permitted, but no bicycles, skates, or skateboards on sidewalks. Each city would have organic food production areas scaled to the population. Open space at the city's edge could serve as field and forest for exercise and renewal, and greenhouses/fields for food production. Cities would be oriented for the maximum generation of its own electricity, and all rooftops and parking areas covered with photovoltaics. Parking garages on the city perimeter could stock publicly shared electric vehicles, fully charged and ready for service. Each city could be connected by local electric rail to main lines of high-speed electric rail. On-site sewage systems would recycle all organics, culture algae for biofuels, collect methane to power backup generators and heaters in the greenhouses, and release only pure water into wetlands absorption areas. Cities might specialize by purpose: universities and colleges, manufacturing, food processing, recycling, retirement, industry, publishing, research and development, corporate headquarters, with government distributed among all.

New Land Use Plan: High-density living would free more land for open space, forestry, and agriculture. More land will be set aside for public use—hiking, camping, water sports. A July 2019 *Science Magazine* article calculated that planting 2.25 billion acres of trees could remove 200 of the 300 gigatons of carbon we have added to the atmosphere since 1800. Excess roads on public lands might be returned to nature. All cities bordering on water could dedicate waterfront to public usage. Mangrove and wetlands can be restored to their former character. No fossil fuels extraction or mining operations on public lands. One goal could be to double lands designated as wilderness by 2050.

Earth Is For Life and Living

"We are returning to our native place after a long absence, meeting once again with our kin in the Earth community. For too long we have been away somewhere, entranced with our industrial world...where we race back and forth in continual frenzy." — Fr. Thomas Berry

Although Father Berry did not cite Thomas Kuhn in his bibliography for *The Dream of the Earth* (1988), he echoes the same conclusion— "The deepest crises experienced by any society are those moments of change when the story becomes inadequate for meeting the survival demands of a present situation."[16] Global civilization now stands on the threshold of the greatest crisis of human history ... ever. The old story has been dying for nearly 300-years and is now on its deathbed. Meanwhile, visionaries from Blake to Berry have been offering both criticism and alternatives. Many are the voices today, calling out like the prophets of yore, "Repent, repent, for the hour of judgement (Gk.-krisis) is near."

Although the word "repent" carries a lot of fundamentalist baggage, it is a most therapeutic word. Literally translated from the Greek, *metanoia*= "another, or higher mind." To repent is to enter a more elevated mind, to gain a broader and wiser point of view, to see the potential for damage and turn away. Poets, preachers, and ecologists have, for centuries now, been standing before the precipice warning Wile E. Coyote as he raced along that he was about to overshoot his management capabilities. Selfish purpose blinds us to risk.

For more than 10,000 years, since the beginning of the agricultural revolution, we have put ourselves in the harnesses of Sisyphus, ever straining to roll that stone to the hilltop, only to repeat, repeat, repeat ... The illusion is that having invested all that energy, we will be able to rest with abundance at the top of the hill—we will be able to relax and profit from the momentum of our investment. But we can never quite reach that dangling carrot, grasp the stars, or overcome gravity that ever draws the stone back to the valley below. Over and over, with new resolve, we continue to try...and continue to waste precious life by a futile bondage to an illusion.

This was the promise of agriculture—settle in one place and farm the land intensely so as to lay up an abundance of food for winter and hard times. But in the process, we needed more people to do the work and gained only more mouths to feed from the same land that became poorer with each harvest. That drove us to acquire more land, to fight more wars, to capture and enslave more people, to wear out more land in a never-ending spiral of enslavement and impoverishment.

More. Better. Faster. These became the guiding principles of civilization. We did not get kicked out of Eden. We chose to leave for the illusion of an abundance that would make life easier, better, and richer. All of this was predicated on the sirens of new technologies.

From there we went on to more advanced technologies—copper and iron smelting and forging, glass, bigger architectural structures that stood as testament to our intellectual prowess. By fits and starts we finally arrived at the Industrial Revolution with all of its promises of ease and abundance. Our new slaves—machines powered by fossil fuels and waterpower—would work for us day and night producing everything the heart could need or imagine. But it was a hollow and deceptive promise. The implied hope was that, once and for all, we could slip the harnesses onto the machine while we enjoyed the "good life."

But that never seemed to materialize. On the contrary, our harnesses only grew tighter and thicker. Instead of having our time free to sit by the river and fish, we had to work ever more hours, feeding and laboring by the machine until we became less human, almost Morlock in nature. The promise of Watt's steam engine turned out to be little more than a Pied Piper leading us to the Dickensian bondage of coal mines, sweat-shop factories full of child laborers, smoke-filled cities, resource wars, the genocide of aboriginal peoples, and massive financial inequity that led to revolutions, inhuman treatment of the poor, environmental destruction, and wage slavery.

This is the fallacy of the growth/progress/advanced technology/ get ahead illusion that we bought into 10,000 years ago and are still laboring under today. What does it matter that we have nuclear weapons? Terrorists will have them sooner or later, also. How will our cyber weapons provide security for more than a few months until nations with 10% of our GDP have them also? We have put ourselves on a treadmill chasing a carrot we will never grasp because it is an illusion. And for 300 years we have been fueling this treadmill by a Faustian bargain with fossil fuels that is now coming due in the form of global warming and massive population numbers that could totally wipe-out all of the supposed benefits we have gained, setting us back to the dark ages.

The questions we need to ask are: Why are we here? What is the purpose of life? What is the best use of life?

Aristotle wrestled mightily with these questions. He concluded that there are three potentials before each man: The life of striving for material gain, which he considered no better than a life fit for cows. Or a life in politics that helps guide the ship of state to improve the lot of its peoples. This he considered an honorable effort. But the highest possible life, he wrote, was contemplation of the eternal order of the cosmos by which a man comes closest to divinity.

The questions of life cannot be divorced from considerations of Earth, since both are entwined. What is Earth for? What are cities for? What is labor for? Are not all for life and the development of the fullest potential of life?

Consider the Gospel question: "What does it profit a man if he gains the whole world and loses his own soul?" Is life just about getting and acquiring more stuff until there is no room left in the house for us? Is it for growing so much food that we kill off the life of our rivers and seas? Is it for reproducing offspring until people are fighting each other to get the remaining resources like a bunch of Yahoos? When will we ever say, "Enough is enough?"

What do you really want? What is the ultimate best use of the life that is now yours? Here, take my wife's kayak and paddle with me to a quiet place far from the madding crowd ...

After a long paddle, we're sitting on a soft sand beach, waves lapping gently at our bare feet, splashing against ancient blocks of basalt. Summer zephyrs, gossiping among pine branches above, carry news from the shores of Andromeda. A ruffled blue waterway lies ahead, dividing two islands, with a golden path dancing from these golden sands to a golden sun raining benedictory photons on these last seconds of a summer's golden day. To live fully in a moment like this is to lay claim to endless time, for moments are the building blocks of eternity.

St. Thomas Aquinas suggested that the end of all our striving was "to become like unto God." Aristotle's God was the eternal order of this world, and the human mind, be it ever so humble, so long as it is focused on this eternal order, becomes one with the God it contemplates. The individual loses her individuality and becomes one with Universal Deity. Thus, to live fully in time, contemplating an eternity that breaks on the shore of each moment, is to fuse with the immortal. To be "religious" is to be "re-ligamented" with the eternal order of the universe.

The past is set in stone, carved with lessons for today. The future is an illusion that never comes. All we have is this moment, which we are ever prone to destroy with regrets over the past or fears of a future we cannot know. All we can do to insure good for tomorrow is live for goodness today. If we fear hateful division tomorrow, let us practice unifying love now. When today is gone, the chance is gone. Life only happens in moments. Lost moments are lost life. Let us be the freedom we want to see. Remember the rich fool—the big-barn builder who stored years' worth of wealth to care for a soul that wouldn't last the night?

Ever and always, life exists on the knife-edge of chaos. On this thin crust of earth, we walk a balance beam, teetering above a molten abyss that would vaporize us in seconds and the icy void of eternal night above. What is to be done must be done in these short moments in the sun, before we, too, become grass on the hillside. Let us learn with Blake, "To see a world in a grain of sand… And eternity in an hour."

And should we not love the riches of time more than the means to riches which most spend an inordinate amount of time worrying about? John Maynard Keynes understood this, "The love of money as a possession…will be recognized for what it is: A somewhat disgusting morbidity, one of those semi-criminal, semi-pathological propensities which one hands over with a shudder to the specialists in mental disease."[17] Consider the Bushmen of S. Africa who work 15 hours/week securing their basic needs and were called "the original affluent society," by professor Marshall Sahlins of the Univ. of Michigan. They are a people unconcerned with material wealth, live in harmony with nature, exist in a social order that is basically egalitarian, uncomplicated, free, and have continued for thousands of years without destroying the land that keeps them.

Hear these words of Stoic philosopher Marcus Aurelius (121-180 AD), "Even if you're going to live 3000 more years…remember: You cannot lose another life than the one you're living now, or live another than the one you're losing…The present is the same for everyone… For you can't lose either the past or the future; how could you lose what you don't have?" Or, Wordsworth, "Getting and spending we lay waste our powers, little we see in nature that is ours. We have given our hearts away, a sordid boon."

In his wisdom, T. S. Eliot counsels us to release the past and live in the present, "Every ending is a beginning. The end is where you start from." Every new story begins where the old story ends.

14. The Last Enemy

"This is eternal bliss, I thought. This cannot be described;
it is far too wonderful!"
— Carl Jung's reaction to a nearly fatal heart attack.
"I seek only the learning that treats of the knowledge of myself and
instructs me how to die well and live well."
— Montaigne
"The last enemy to be destroyed is death."
— St. Paul

The Legs of "Ozymandias"

LEGS HAVE ALWAYS BEEN AN ANALOGY for strength and power. The muscles of the thighs are the strongest in our body. They enable us to stand, run, and conquer. Likewise, ideologies have legs that both empower and hold them erect. Tyrants, empires, and ideologues have always constructed towers—symbolic of supporting legs—as visible metaphors for their power and virility. When legs give out, both bodies and ideologies come tumbling down. This is the deeper significance behind the demolition of the twin towers of the World Trade Center on 9/11. Osama bin Laden was signaling the demise of Western power.

The paradigm that is currently disintegrating has stood for 10,000 years on three legs. One is the leg of **dualism**—the separation of God from Earth, soul from body, light from dark, us from them, life from death. The second is **dominionism** that believes Earth was given as a storehouse for man's needs, and man--the image and likeness of God— has a divine mandate to use those resources as he sees fit. And the third is the fear of **death**. People who fear death are easier to control. They are pliable and obedient to authority that (supposedly) will protect them from death or provide a way through death to life eternal. The author of the *Book of Hebrews* (perhaps St. Paul) understood this when he wrote that (Jesus) "by his death might destroy him who holds the power of death—that is, the devil—and free those who all their lives were held in slavery by their fear of death" (Hebrews 2: 14-15). The power of death is a lie, from the father of lies. Fear of death issues in servitude. The person who knows the truth will not be enslaved.

Interpreted within the framework of Norbert Wiener's cybernetics and his Manichaean Devil, this quotation holds a deep insight. According to Wiener, the goal of the Manichaean Devil is to consciously lie, disrupt, and jam messages in ways that are systemically harmful. From a system's point of view, breakdown in the flow of information is death, because systems run on messages. When messaging stops, vitality ceases. Because systems are minds, and minds feed on information, stopping the flow of information is equivalent to starvation—they run out of gas. Fear of death is a powerful motivator. In terms of Spinoza's *conatus*, death is the archenemy of "continuance." Therefore, systems are especially alert to messages of death and dissolution and do all in their power to exercise avoidance behavior.

But what if the concept of "death," as commonly understood in Western Culture, is one of the biggest lies to jam message that has ever come down the road? By that I mean, what if this idea of cessation of being, the dying of "self," the end of life, is just a super-big hoax that has been employed for millennia as a means of power and control over naive masses of men?

What if the threat of death has become one big, powerful tool for manipulation, extraction, and entitlement for free riders? Barbara Tuchman makes a credible case for this in *A Distant Mirror* (1978), where she presents abundant documentation regarding the depredations of the Medieval Church, leveraging the fear of death and hell to extract monies, while upper level leadership lived amid opulent excess.[1] Why else would the economic elite seek to align themselves with a religious ideology they personally do not believe? If they believed, would they not seek, by word and behavior, to be examples of the tenets of that religion by acts of compassion, truth, kindness, forgiveness, love?

Furthermore, if death occurs when information ceases to flow, then many religions cultivate death to survive. Dualistic religions tell adherents that God is mysterious, and we cannot understand his ways. We cannot understand the relatedness of all life, because creation took place as a miraculous act by divine fiat. Therefore, the world is unintelligible because there is no linkage with the greater universe, no understanding of process, and no derivative relationships between life forms because each was created as a distinct, stand-alone, being. Heaven, hell, a transcendent God, creation by divine fiat, miracles that defy the laws of physics—all break the flow of information and deny any rational basis for understanding. Ideologies protect and justify unintelligibility for the same reason Jesus said, "Men love darkness more than light, because their deeds are evil." Once a light shines under the cabinets, the cockroaches lose access to free food.

This is why a ruling ideology resists the concepts of biological evolution that demonstrate our linkage with all of life, ecological truth that provides clear causal relationship between human sickness and industrial pollution, seeks to protect dark tax havens and shell corporations, and funds disinformation to discredit the proven science behind global warming. Establishment of causality is equivalent to shining a bright light on sources that issues in liability. The fossil fuel industry has desperately fought causality for global warming. Fundamentalist religion crumbles before the science of evolution. Patriarchy is threatened by equality of the sexes. Nationalist politicians will never accept racial equality.

But the biggest bastion of darkness is the fear of death. It is the third leg that holds Ozymandias erect. Once the clear light of day shines on the illusion of death, this idolatrous ideology that has kept mankind groveling in fear, guilt, shame, and servitude will finally crumble back into the common sand from which it was made. In order to maintain their power, ideologues must keep death alive. If death dies, they lose their leverage.

What Is Death?

By now some readers may be shaking their heads, thinking I am attempting to deny the obvious. Yes, it is obvious that the bodies of living things die and eventually disintegrate. This can be desirable. Friends had us in for dinner last week. Our glasses were kept filled with wine amid stimulating conversation. Swordfish was on the grill, and it smelled delightful. Technically, that swordfish was dead. But we were all quite happy about that, for death meant life and communion to us. By the end of the meal, not a scrap remained on the platter. We had each taken part in the ritual of shared food that goes all the way back in time, one that bound us in life-supporting groups before there was any formal religion or culture. I left feeling sustained in body and mind.

On the other hand, death is often ugly, repulsive, and putrid. A porcupine waddled out in front of a car the other day and got squashed. For obvious reasons, nobody wants to stop and drag a porcupine off the road, so it has remained there for days. It is now maggot infested, and stinks to high heaven. Drivers avoid contact. Eventually, it will be beaten down and disintegrate enough for a good rainstorm to wash the remnants off the road.

So, yes, death is a reality in this world. Bodies get dressed, perfumed, and lie in funeral homes for a day or so until they are lowered into the Earth. Others are cremated and spread under trees or on great waters. The facts are beyond dispute.

But I do not think that is what we really fear when we fear death. Ultimately, the fear of death is the fear of finality, of non-continuance, of lost relationships, of powerlessness, and the vanishing of "me." Death is the end of our story, bathed in a massive sense of futility. And for those who remain, death punches a big hole in their lives by the loss of a loved one. But if we dig down deep, to bedrock, I believe that in dreading death we are resisting the loss of unique informational patterns.

If the driving force of life is continuance, cessation of our story is a threat to continuance and issues in feelings of negative valence because it raises the probability that our story will come to an end. This sets off the fire alarms within patterned control systems that have been programmed from the beginning of time to strive for continuance. Since all of life and evolution is an effort toward increasing the complexity of syntactically patterned information, the promise of de-patterning is an assault on meaning. Death is a mocking face rising up to pronounce futility to our illusory self. This is the ultimate despair of dualism, because here at the end of life we are left with a heaven that was nothing but a hope and a fading Earth to which we never belonged.

This leaves us in a really bad place. On the one hand, all of the struggles, sacrifices, accomplishments, fears, love, and joys will just be dumped into a hole in an Earth that, according to dualistic religion, is due to pass away in a denouement of futility. On the other hand, we are exhorted to have faith and trust, confident that the transcendent God of the universe will descend in human form, riding a white charger, to wage war on the very devil and demons he must have made in the first place, and consign them to a lake of fire and brimstone, along with the innumerable billions of humans, made in his own image and likeness, who will likewise be assigned to everlasting torture worse than that of Prometheus.

I see only two possible interpretations of this scenario: Either it is a myth (full of deep truth) intended as a mirror reflecting to us the logical end of our selfish, conscious, linear purposes that are sowing both social and ecological havoc on Earth and can only end in some apocalyptic finale. Or God is a sadist who is the diametrical opposite of the love, compassion, and forgiveness exemplified in the person of Jesus, who is identified as his son. I go with the first option because the second is totally unintelligible—sicker than Charles Manson.

Please, do not pass over this lightly. It is the gasoline fueling the fires of global conflict today—us against them, believers against apostates, heaven against hell, God struggling with Satan, dominionism versus sustainability. And it is driven by the fear of death and longing for immortality. Perhaps fear of death is just an expression of ignorance that appeals to our reptilian brain and is leveraged against us. Unintelligibility leaves us vulnerable.

But what if death became intelligible, and was addressed by the cerebral brain? Granted, death, like consciousness, is a hard problem. But that is no reason to throw our hands up and walk away. Cognitive scientists still cannot fully explain consciousness, but they do know a whole lot more than 20 years ago and are making steady progress all the time. Just the same, we probably will never know all the mysteries surrounding death, but we are making progress and getting glimpses behind the curtain. And those glimpses, many anecdotal, are very encouraging. For one, refer back to the quote from Carl Jung, one of the greatest psychologists of the 20th century. It is a subjective experience, but I see no reason for Jung to report other than his true feelings.

Now, just suppose that the Abrahamic religions have the story backwards. What if it is the soul (our self-story) that actually dies, while the body, via its material substance and genetic coding, lives on into an undetermined future? This is the more cerebral option because we know that the atoms and molecules that compose this physical body will not just disappear but continue on as long as the universe exists. And our coding has continued in an unbroken chain since proto-bacteria and will go onward as long as life exists. Therefore, we are already participants in eternal life, regardless of belief or unbelief.

Experience, Myth, and Opinion

Religion must be regarded as the ultimate longing for *conatus*—the striving to continue. Most religions have some story to account for the post-death continuance of a person's essence. That essence is variously called soul, spirit, or mind—all immaterial essences of our being. Across the gamut of cultural myths, these are symbolized by the invisible forces of nature—wind, fire, breath, fog drifting across a marsh at dusk, etc.

Gregory Bateson, one of the founders of cybernetics, participated in many of the Macy Conferences with his then wife, Margaret Mead. He was an atheist, yet near the end of his life came to believe that only a "religious" response, motivated by the aesthetic instinct, could save us from an ecological catastrophe. He understood that our self-centered, conscious linear purpose, ignorant of nature's vast network of interconnected patterns, could, when coupled with the multiplying powers of technology, inflict destruction beyond our capacity to respond. Concluding his Korzybski lecture in 1970, ten years before his death, he said, "At last, there is death. It is understandable that, in a civilization which separates mind from body, we should either try to forget death or make mythologies about the survival of transcendent mind. But if mind is immanent ... then death takes on a different aspect. The individual nexus of pathways which I call 'me' is no longer so precious because that nexus is only part of a larger mind."[2]

For Heraclitus, psyche (soul or mind) is the vital force that indwells the cosmos and is the same as the vital force that indwells humans. Euripides held to the natural view that at death the individual soul reunites with the immortal world soul. Plato is more concerned with the act of dying rather than death—a noble death validates the life of the philosopher and is of no concern since it only affects the body. For the Stoics, death is the test of right thought and conduct, and serves as the positive purpose that gives meaning to life. Men who do not value contemplation are regarded as dead while they live.[3]

In a previous chapter, I discussed human transcendent experiences. These were of three divergent types—deep, personal religious experience as documented by Wm. James, the experience of psychedelic drugs documented by Michael Pollan, and near-death experiences widely documented in current literature. First, it is important to say that these are mostly subjective, anecdotal accounts. Likewise, it is clear that though individual temperaments, times, situations, and backgrounds are widely divergent, there are areas of marked commonality to all of these experiences. In addition, scientific rigor was imposed on some of the psychedelic experiences done by researchers at Johns Hopkins University and New York University.

One of the universals in all three classes of transcendental experience is the loss of the fear of death. A few months ago, my wife and I got together for dinner with new friends before attending the presentation of *Funny Girl* by a local theatre group. Leo was friendly, but tastefully reserved. At least until my wife mentioned this book I'm writing. I said that it was about a *New Story*, a new way of looking at life, relationships, religion, culture, and death. At that, Leo perked-up, "I almost died and had this amazing experience." He proceeded to describe what thousands of people have described as the experience of dying and merging with a unifying light that is pure love. After he had finished, I asked him one simple question: "Leo, how has this experience changed your view of death?" Without a moment's hesitation he replied, "I am absolutely and totally unafraid to die. Death is a beautiful thing. There is nothing to fear."

Wm. James documented the exact same reactions, as does Michael Pollan. Invariably, people experience a oneness with all of time and nature, human consciousness merges with the leaves of the trees, clouds in the sky, the grand pageant of evolution across eons of time. This experience of ultimate connection with all things can only be described as a deeply religious experience, even though many of the subjects remained self-proclaimed atheists.

This is of great concern to organized religion because the keys to unlock the fear of death are no longer the sole domain of sectarian systems. People of all religious backgrounds and denominations have convergent experience and lose their fear of death by connection with the greater universe.

Nothing much has changed down to the present day. Neither the government nor the Church wants people to lose their fear of death, because people who are unafraid to die are less willing to subject themselves to an authority weaker than they have already experienced. Why would we want to kill others with whom we are one, or destroy the natural order of which we are an integral part, invest in the cult of celebrity and consumption, or endure the unintelligibility of religion when we have had first-hand experience of the "sacred"? And how does this happen?

There is no clear answer, but cognitive scientists believe it has to do with the shutting down or overriding of the Default Mode Network (DMN) that is like the conscious clearing house for mental operations to keep us alert and alive by prioritizing current experience over subconscious message.[4] Whenever the DMN goes off-line, other areas of the brain get to inject their two-cents worth. This happens during the three above mentioned experiences, plus during dreamtime when subconsciousness is permitted to speak. At these times, the many areas of the brain suppressed by the DMN are allowed to chat directly with each other. In the process, they reveal an entire universe of reality and connections that we are totally unaware of in conscious states. It is this deeper reality of infinite connections that binds us in a unity with all of time, life, and the universe. And it is beautiful beyond description. The blood flowing in our veins merges with the sap rising in trees. Clouds flow into our lungs and then ride out over mountain ranges. We experience God in the smallest flower and the farthest star and in a love that enfolds us more completely than our own skin. What experience or liturgy does institutional religion offer that comes even remotely close to this—that would motivate us to care as deeply for our fellows and the natural world, and accept death as just another step in the process of being?

Stump Preachers

"If a man dies, shall he live again?" — Job

Roughly 35 years ago, a developer began carving this land into parcels for new homes. Either he, or the seller, selectively logged the area at that time. Remnants of old stumps remain, along with traces of skidways used to drag logs to a loading platform. Sections of barbed wire still hang, rusting and unconnected, embedded in tree trunks at the wooded edge of former pasture. But it's the stumps that speak a greater truth to the woodland walker.

The day they were cut, bacteria and fungus invaded the bleeding wood and began to convert cellulose into humus. After a few seasons, moss and lichen took hold and increased the rate of humus production. Then, one fine day, small seeds of pine and hemlock, wafting on a vagrant breeze, landed in the mosses. Come spring, they sprouted and began to grow. And now, lo, these many years passed, they are two feet tall, erect, green, their feet adorned with moss and lichen and teaberry, standing strong in their place like little "Stump Preachers" proclaiming their own gospel of resurrection and life unending. They don't confront me with questions or threats, a creed, recursive philosophical arguments, or the need to join an organization. They don't yell and pound their sylvan pulpits. Theirs is a simple catechism, (LTC): Look closely; Think broadly; Connect the dots. Centuries of leaf fall, acorns, frogs, and deer that decayed have returned once more as life...and will return again and again into ages hence. Their sermon is both simple and profound: "We are all immortal, changing and flowing one into the other in a never-ending circle. We are of one substance, connected in a web of eternal being."

In the larger view (the rational, cerebral view), nothing in this forest is dead. Fallen trees melt into humus, iron from old, barbed wire will end up in a squirrel's hemoglobin, stones become micronutrients— they are all in process, being changed from one thing into another. Each particle is like a pendulum swinging from one body to the next, always striving for equilibrium but never stopping. To stop, to remain inert at full equilibrium, to resist change, is to embrace true death— separation. What we erroneously call death is merely a new beginning, an adventure in standing up again. Wallace Stevens referred to death as, "The mother of beauty." My Stump Preachers merely laugh at the idea of death and ridicule it by their simple, stand-up again, presence. "Hey man," they often say as I walk by, "look about and see, you are immersed in immortality."

Dying of the Light

Summer is closing down. Maples around the edge of the pond are bursting forth in color—saffron, salmon, red, and gold. The land is hushed, as though waiting for a change of the guard. Fog lingers later in the mornings and comes earlier at dusk. The geese are growing restless, knowing their time at the pond is short. Frogs are quiet, and all the water lilies are gone. Change is in the air. I am thinking of Dylan Thomas, "Rage, rage against the dying of the light." Come, sit with me for a while to embrace this moment.

Days grow shorter. Light comes later to the mountains. There is a portent in the air, something not seen but sensed, like a vagrant hint of winter riding on lowering light. The sun rises slowly in an ethereal opalescence flooding across the hazy marsh to become a living force, glowing from within the fog. Tree trunks warm and steam. Moss-covered rocks shimmer like massive emeralds. Sunlight enters the cabin and spills across the wide floorboards. (Do they remember the sweet blood of a tree?)

Eventually, the magic of morning passes into the light of common day. Fog banks have galloped away like ghost horses before a rending sun. Birds grow silent. Flowers lift their happy faces. The wash is hung to dry. Mowers and motorboats hum less often in drowsy September. The cares of life follow their mundane and plodding paths. Morning morphs into afternoon; afternoon, at length, distills into the last light of a fading day.

Eventually, Apollo's horses grow weary. Downward they plunge, drawing the fiery furnace of creation to the cool of evening, the cool of the sea, where they will descend to rest at the center of Earth.

In these last few moments of an aging day and season, the sun pauses to cast a glowing benediction over Earth. Long shadows stretch across the land. Ridges catch at the last light as evening slowly rises from valley to peak. In these closing minutes the open space above marsh and meadow is transformed into an aura, living and liquid, an amorphous and generative medium. All that was to be, has been. The day's work is done, the harnesses hung. The nail is hammered; the hinges oiled; the door is swinging; the catch will close.

Earth and man can no more rage against the dying of the light than one can rage against the mother who gave the precious gift of life. Come, sit with me here on this granite slab beneath the hemlocks and maples and let us, in these last sacred moments of day, learn to embrace the gentle night. See them now, rising upward on the last warming currents—insects and swallows take wing. Circling, darting, diving they scribe ambiguous arcs in a final dance with the sun. Bathed in a bronze light, they become translucent grace drifting to and fro on gossamer wings. How else to interpret this, but as a celebration and surrender of gift to the giver from which it came? Surely, only the confused, a thrashing and sad soul—the spawn of falsehoods—could stand on the threshold of life's great commencement and rail against eternal grace. (What is there to fear more than twisted thoughts and words?)

At the end of a long and wearisome journey, William Clark watched the sun setting over the Pacific and wrote, "O! The joy!" The

only reasonable response at journey's end is to celebrate the dying of
the light, to recite that final *Eucharisteo*—"I give thanks ...O, The joy
..." In a few moments, our swallows and dragonflies and midges will
have spent their final ecstatic moment in the sun and then descend
downward to Earth, downward to death on wings flashing the last
golden light, wings of thanksgiving. They show us how to live ... and
how to die ...

While a soldier in Iraq, fighting both an absurd war and the terror
of death, Roy Scranton sought counsel in the *Hagakure,* that counsels
the 18th century Samurai, "Meditation on inevitable death should be
performed daily."[5] Using this as a steppingstone to a larger argument,
Scranton suggests that the time has come for us to release this dying,
carbon-fueled dualist civilization to the inevitability of death. People
die, and ideologies die. The curtain must soon fall on this old act,
before the next new thing can take the stage. The new can begin only
at the end of the old.

It seems, therefore, that there is no such thing as a cessation of being we
call death. Perhaps what we call death is merely like a wave thundering
its way down a beach. It rises, builds, curls in its most elegant form, then
crashes down only to be reabsorbed by the greater sea. After a while
it will come again and again and again. But it is not the same wave
as before. Moreover, it has merged with all the other molecules of the
sea and all of the other "once-waves" to become something totally new
yet ever the same, as it rises, curls, and crashes onward into eternity. It
may be the finest wave you have ever seen. But it is the same as all the
previous waves that have enchanted us and our relatives for 2,000,000
years. Ever it lives. Ever changed. Always, it is the same—immortal.

What Will It Be Like?

The time has come for death to die. When death dies, fear will die. When fear dies, freedom is born. Without death, there will be no birth. Death and birth are both painful struggles. Birth feels like death, and death feels like birth.

It seems to me that the world of this present moment is enduring increasingly sharp labor pains. The old order is dead, but the stinking body is still cluttering the stage and those who profited from it refuse to let truth carry it off for a decent burial. They are lashing out at the ones who want to move forward to a new world with a new way of thinking, being, and living. Their dedication to the old body is so intense that one must wonder if they, too, must die before we can clear the stage and scrub the floors for the new act. It has been this way in every age. To some people, change is personal death, even when it becomes quite clear that failure to change only assures death. We are sometimes so selfish that we would take others down with us to defend an irrational point. Dying and flailing dinosaurs are dangerous.

Perhaps it doesn't matter all that much. Life will find a way—always has and always will. No individual can save a civilization bent on destruction. All we can do is live our lives and be the change we want to see. A major part of that change is to learn to die with grace and joyful acceptance. For, assuredly, we must die to be born anew on some far-off day in spring, when daffodils are pushing up from thawing Earth and stars sing in the heavens.

Do you wonder what your death will be like? I do. And I think it will be an ecstatic surprise—a bliss too wonderful to describe. Countless thousands of people have testified to a state of transcendent euphoria derived from a brush with death. Nobody paid them to lie.

What Will It Be Like?

What will it be like at the moment of death?
When we with a sigh give up our last breath?
They say when we die
The light leaves our eye.
Where does it go?
Don't you want to know?

To be continued…

15. The Wrap

Assessment

AND NOW, THE BIG QUESTION: What are the probabilities for the birth of this new paradigm? I've given this a great deal of thought. My conclusion is that this new paradigm has a 100% probability for realization. Although the form may, and probably will, differ from what I'm seeing, the basic outlines will come to pass. This is good news.

The bad news: I find scant evidence that mankind has, willingly and without great struggle, chosen to alter ingrained behavior because of some future threat. We seem programmed to push ahead with an agenda that has worked well enough for hundreds of years and brought not a few benefits. Our addiction to fossil fuels is but one example. The bigger problem, however, is that we often choose complacency because change is a threat of assumed loss. Since politicians derive their power from the elite, and the masses are so easily deceived through messaging crafted by the elite, the necessary political change will probably come too late. Hegel's *Owl of Minerva* will have flown. We may exceed the tipping point for apocalypse.

Economic and religious interests will fight against any campaign to reduce birth rate. We need to get back to a human population nearer to 3 billion to be in balance with the Earth's natural carrying capacity. That's a loss of 4.7 billion from current levels. If every woman had but one child, we could be there within a generation. But that will not happen, particularly in the developing world where ignorance, corrupt governance, and fundamentalist religions hold sway.

Sooner or later, someone is going to let go with nuclear weapons—perhaps suicidal thermonuclear devices salted with cobalt if it is a poor nation unable to withstand the onslaught of a powerful nation (Jonestown on steroids). Decreasing food supplies at a time of increased population will make immigration pressures unbearable. Desperate people do desperate things. There will be blood.

Nevertheless, I remain a long-term optimist. Life found a way to colonize Earth billions of years ago. That was no mean feat. RNA had not yet been invented. It was hot. The atmosphere would have killed us. And there was no pizza. Yet, life built a beach head, evolved, changed the environment in ways that fostered new varieties of living things, greened the Earth with vegetation, broke down rock to form soils, and marched onward to us.

But not so fast. During this process there were multiple reversals, die-outs, ice ages, volcanoes, and evolutionary dead-ends. And yet, and yet … life marched forward, sometimes falteringly, but ultimately onward and upward. Eventually, the Earth was full of beautiful and complex forms of life—plants, insects, invertebrates, and vertebrates—all ranging over land and sea carving out niches for themselves and their progeny. Earth—this once hot, toxic, inhospitable rock floating in space—became home, a place of verdant mountains, indigo-colored seas, thundering waves wearing down the hem of continents, massive flying serpents, bellowing dinosaurs, and finally this strange primate with a big head who discovered how to use fire and cook food and leave his red-ochered dreams on the walls of caves.

Already But Not Yet

Can an individual find personal peace in the midst of a massive vortex of daily news that would suck the goodness out of life? How can one carry around this weight of awareness that is our responsibility as thinking adults, and at the same time face life with sincere optimism? Can we stare down an apocalypse while carving out our own little Eden; live under a corrupt political regime, call out its depredations, while building islands of wholeness?

If Montaigne and the philosophers are correct, healing begins with the long view. Civilization gave us science, technology, medicine, and the arts. But underneath all the advances has been an unending striving for power and wealth that often made us into less than animals by our violence, wars, genocides, and present ecocide. Viewed in Earth-time, however, our destructive illusions are simply random trials that are maladaptive and will be deselected by Nature. This has happened millions of times before. If certain traits do not align with environmental parameters, they do not survive. Ultimately, systems evolve to fit. Those that fit, survive to reproduce and continue. The takeaway is simple—the Logic of Nature, not man, is the ultimate selector. Nature (God) always votes last.

What we resist, we empower. The zealots who fought Rome, empowered Rome. Those who fight the fossil fuel industry merely empower it. Those who hurl expletives at politicians, empower politicians. We kill bad ideas not by fighting them, but by replacing them. In other words, we must come up with a better narrative before we can move forward into freedom. As individuals begin to live by a better narrative, there will be an alternative as Ozymandias' legs crumble. The world needs a post-apocalyptic option. If we fight death, we will begin to look like death. Better to let the dead bury the dead and point to the light ...

Here is an example of how it worked for me yesterday: I had been writing all morning about our destructive falsehoods and their effects on both Earth and man. It's hard to keep this stuff from seeping into one's core and painting the halls black. I needed to come up for air and light.

So, I took the kayak down to 20-Mile Bay and launched into a steady, but light, drizzle. I enjoy paddling in the rain. The wind seldom blows, the surface of the lake is calm, and the paddling is easy. Moreover, lowering clouds and the absence of motorboats creates an intimate atmosphere where I feel even more connected to the water than normal.

Taking my strokes long and easy, I made my way past Spectacle Islands, where loons had a nest on the beach. A little further on, just offshore from Nine Acre Island, I came across the mother loon with her one chick sitting on her back. The mother knows me because I give an imitation loon whistle when I see her. She appeared completely at ease.

Gliding silently along the western shore of Nine Acre Island, I noticed a pair of large black eyes peering at me from the brush above the shoreline. It was a deer arrayed in her reddish summer coat, ears full forward, and nostrils flaring. Her head was totally encircled by lush green foliage, as though framed for a rustic picture. In that instant I had the distinct impression of looking at ... myself. There seemed to be a silent knowing going on between us, one studying the other as a mere variant separated only by time. There was a connection. For but a brief moment, we were one in communion. A heaviness lifted from my heart. There was assurance that life will go on. Things will be OK. The new story is already here in the world of Nature, but the old story has not yet defeated itself.

The new story is visibly among us, but we have to choose it. To choose one is to release the other. Loyalty to both the new and old order is division which results in instability. The heart thrives on unified allegiance. Both the temporal order of cultural memes and the eternal order of nature coexist. One will pass, the other will continue. We can choose our allegiances. It may not be easy, but it is possible.

I think this is what John Steinbeck was trying to tell us ...

Timshel

In 1952, John Steinbeck published his epic *East of Eden*, chronicling the intertwined lives of two families—the Hamiltons were poor Irish immigrants, and the Trasks, because of a large inheritance, were quite rich. Set in the valley of the Salinas River of California, the Hamiltons carry on in their salt-of-the-Earth way, trying to scratch a living out of a hardscrabble piece of earth. The Adam Trask family, however, has abundant resources, a fine piece of land, and a Chinese cook and houseman named Lee. Adam is a kind and generous man who had the great misfortune to fall in love with an evil woman named Kate. Kate had murdered people with no regret, was a prostitute, and a selfishly evil schemer. Adam had twin sons by Kate (although he was probably not the father)—Aron, who is sensitive and kind, and Cal, who is a schemer like his mother.

As it turns out, Lee is quite the scholar and spends time with an elite group of Chinese intellectuals in San Francisco who had learned

Hebrew from a rabbi. Talking one day with Samuel Hamilton about the story of Cain and Abel in the fourth chapter of Genesis, Lee recounts how his group had discussed the story for two years before deciding on the meaning of one Hebrew word.

Cain was angry with Abel because God favored Abel's sacrifice. God counseled Cain to drop his evil desires and do the right thing. Although the urge may be strong, a man can control his evil impulses, change his outlook and ways, and make life-enhancing choices, "If you do what is right, will you not be accepted? But if you do not do what is right, sin is crouching at your door; it desires to have you, but <u>you must</u> master it." —Gen. 4:7 (NIV)

Steinbeck's entire novel rests on this one Hebrew word—*timshel*—which he interprets, "thou mayest." The New International Version quoted above interprets *timshel* as "you must" master it. The clear message is, "by choice, we determine our future lives."

By the book's end, Aron has died in the Army and Adam is on his deathbed. Cal, who has inherited his mother's propensity to evil, is standing beside his father's bed, feeling guilty about his brother's death (Cain and Abel, redux). Lee pleads with the dying Adam to bless his son and set him free to live. Haltingly, Adam struggles and finally utters, "Timshel ..."

Timshel—we may; we can decide; it is in our power to change. We are human and stand in the stature of gods, making a future by choice. We can choose evil. We can abdicate our humanity, cower in fear, and make no choice. Or we can choose a new way—the way of life, the path that leads from Baal and Ozymandias to a new future in a free and good land. We can choose to throw off the yoke of bondage, to reject idols—the falsehoods—girt up our loins and march toward the light of a new day and a new way of living and loving. *We may ...*

And when we make that choice, we will find that we are not and never have been alone. Truth, Light, and Universal Logic is and always has been with us. Deity—the Logic of Nature—is immanent at the center of this world, connecting us in an unbroken web of belonging with all that is, has been, and will be. We are not orphans. Dualism, Dominionism, and Death are lies. This is our home. We belong and we will continue. *We may...*

Coda

IT IS NOW AUTUMN. The maples and birch around the beaver pond are iridescent in October's late afternoon sunlight. Waterlilies have descended back to mud leaving the pond one unbroken cobalt surface that reflects each drifting cloud. The geese survived the days of waiting for late winter ice to melt and successfully raised their families. Two dozen goslings are now fully grown and make multiple training flights each day, honking their way around this vast open meadow. A signal will soon come on slanting beams of light and they will set off for their long journey south. Then, one day in December, ice will cover the pond once more and silence will rule the land.

This book, which was little more than a collection of seed ideas in mid-winter, has found its way to final form here in the days of ripening fruit. Thoughts and ideas have their own harvest.

At the outset, I was uncertain, hesitant. How could a man, who for a large part of his adult life proclaimed a message of tribal superiority, Biblical inspiration, the entitlement of the faithful, and a theology of dualism with evangelical fervor, gradually turn 180 degrees in favor of an inclusive, scientific, monistic naturalism that promotes belonging and ridicules death? Was I losing my faith—becoming an apostate? Would I become the once maligned atheist enemy?

In the process of writing I did leave that old God, or that old conception of God. That God became too small, too provincial—a manufactured story with no empirical basis—invented by people for fear and advantage. But I emerged with something of far greater value, as light surpasses darkness.

In the process I understood a bigger God—a God as big as the universe. I discovered a God who is not separated far away in an imagined heaven, but who is present in every logically patterned form of being, from rocks to trees to the vast ecosystems of Earth to the light in another person's eyes. This is the deity in whom we are immersed at every moment, who is reflected to us in every falling leaf and every sunset, who speaks wisdom in the honking of geese preparing now for migration to warmer climes.

And lastly, there was death, that old enemy that sows terror far and wide. But what I once feared, now I embrace with an honest ease. Because we are merely a temporal form of the Logic that rules the world, we need no longer be concerned. The matter that is "us," rendered by the chaos of a future that is both unknown and unknowable, will continue in a multitude of forms yet unimagined. And our core reality—the Logic of the universe—how could that ever be destroyed or nullified? Thus, our essence (not this illusion of separate selves) will continue onward in newer, higher, and more marvelous forms to discover yet again and again that old felt divinity in a moon rising over future mountains, flashing on silvered waters, dancing in the silken threads of a dew-ladened spider's web, ever enfolding us in the enchantment of a soft summer's night.

Endnotes

Introduction

1 Harari, Yuval N. (2018), 21 Lessons for the 21st Century, Spiegel and Grau—Random House p. 263
2 Kuhn, Thomas, (1962), The Structure of Scientific Revolutions, University of Chicago Press, p.150

1. Information

1 Bateson, Gregory (1972), Steps to an Ecology of the Mind, Ballantine, p. 459
2 Wheeler, John A. (1994), in At Home in the Universe, American Institute of Physics, p. 298
3 Dawkins, Richard (1986), The Blind Watchmaker, Norton, p. 112
4 Sagan, Carl (1996), The Demon-Haunted World, Ballantine Books, p. 114
5 Wheeler, John A. Ibid p. 296
6 Lloyd, Seth (2006), Programming the Universe, Knopf, p. 3
7 Lloyd, Seth, Ibid p. 15
8 Lloyd, Seth, Ibid p. 154

2. God is Meta-Logic

1 Kleinknecht, Hermann, (1964), Lego, Vol. 4—Theological Dictionary of the New Testament (TDNT), Eerdmans, ed. by Gerhardt Kittel, trans. by Geoffrey Bromiley, pp. 87-91
2 Birkhoff, George, and von Neumann, John, (1936) The Logic of Quantum Mechanics
3 Quantum Logic, Internet Encyclopedia of Philosophy
4 Kurzweil, Ray (2012), How to Create a Mind, Penguin Books, pp. 82-5
5 Kurzweil, Ibid, p. 90
6 Heisenberg, Werner, (1958), Physics and Philosophy, Ruskin House, London
7 Lyre, Holger, (2003), Time, Quantum Information, Berlin, p. 178

3. Logical Information Patterns

1 Bateson, Gregory, (1979), Mind and Nature, Dutton, p. 10
2 Tooby, John and Cosmides, Leda, (1992), The Adapted Mind: Evolutionary Psychology and the Generation of Culture, Oxford University Press, p. 56
3 Thompson, D'Arcy Wentworth, (1961), On Growth and Form, Cambridge University Press, pp. 326-7
4 Capra, Fritjof and Luisi, Luigi, (2014), The Systems View of Life: A Unifying Vision, Cambridge University Press, pp. 129
5 Miller, Peter, (2010), The Smart Swarm, Penguin, p. 10

308

4. Mind

1   Bateson, Gregory, Mind and Nature

2   Jung, C. G., (1986), The Symbolic Life: Miscellaneous Writings, Collected Works, Princeton University Press

3   Schrenk, Gottlob, (1964) Nous, Vol. 4—TDNT, p. 956

4   Stapp, Henry, (2010), Minds and Values in the Quantum Universe, in Information and the Nature of Reality, ed. by Davies and Gregersen, Cambridge University Press, p. 109

5   Pinker, Steven, (1997), How the Mind Works, Norton, p. 24

5. Causality

1   Deleuze, Giles, (2010), quoted by Bernard, Olaf, and Kruppers in Information and the Nature of Reality, Ibid, p. 182

2   Farnsworth et al., quoted by Farnsworth, Ellis, and Jaeger, Living Through Downward Causation in From Matter to Life, (2017), Cambridge University Press, p. 330

3   Marletto, Chiara, (2017), Beyond Initial Conditions and Laws of Motion: Constructor Theory of Information and Life, in From Matter to Life, Ibid, p. 39

4   Hoffmeyer, Jesper, (2010) Semiotic Freedom: An Emerging Force, in Information and the Nature of Reality, Ibid, p. 201

5   Zenil, Schmidt, and Tegner, (2010), An Algorithmic Software Approach, in From Matter to Life, Ibid, pp. 262-4

6   Carey, Nessa, (2012), The Epigenetics Revolution: How Modern Biology is Rewriting Our Understanding of Genetics, Disease, and Inheritance, Columbia Univ. Press p. 33

7   Neihardt, John, (1932), Black Elk Speaks: Being the Life Story of a Holy Man of the Oglala Sioux, University of Nebraska Press, p. 150

6. Complexity and Chaos

1   Turing, Alan M., (1950), Computing Machinery and Intelligence, *Minds and Machines* 59, no. 236, as quoted by James Gleick, The Information, p. 377

2   Mitchell, Melanie, (2009), Complexity: A Guided Tour, Oxford University Press, p. xiii

3   Ibid, p. 13

4   Wilson, E. O., (1984), Biophilia, Harvard University Press, p. 36

5   Quammen, David, (2018), The Tangled Tree: A Radical New History of Life, Simon and Schuster, pp. 152-5

6   Walker, Brian and Salt, David, (2006), Resilience Thinking, Island Press

7   Gleick, James, (1989), Chaos, Penguin, p. 110

7. Life

1   Kauffman, Stuart, Antichaos and Adaptation in Scientific American, #256 pp. 78-84

2   Ralston, Holmes III, Information and the Nature of Reality, Ibid, p. 223
3   Adami, Christoph, and LaBar, Thomas, (2017) From Entropy to Information, quoted in From Matter to Life, Ibid, p. 8
4   Davies, Paul, (1998), The Origin of Life, Penguin, pp. 40ff
5   Davies, Paul, Universe From Bit, in Information and the Nature of Reality, Ibid, p. 80
6   Capra and Luisi, Systems View of Life, Ibid, p. 236
7   Quammen, David, The Tangled Tree, Ibid
8   Eiseley, Loren, (1964), The Unexpected Universe, Harcourt, Brace, and World, p. 231

8. Truth

1   Wiener, Norbert, Cybernetics and Society, Ibid, p. 130
2   Orwell, George, (1943), from Looking Back on the Spanish War, quoted in Orwell on Truth, 2017, Houghton Mifflin Harcourt, p. 81
3   Bateson, Mary Catherine, (2015), The Illusion of Certainty, contained in This Idea Must Die, ed. John Brockman, Harper Perennial, p. 491
4   Jung, C. G., (1961) Memories, Dreams, and Reflections, Random House, pp. 67-8
5   Grundmann, Walter, Kalos, Vol. 3 TDNT, pp. 536-43
6   Prum, Richard, (2017) The Evolution of Beauty: How Darwin's Forgotten Theory of Mate Choice Shapes the Animal World—and Us, Anchor Books

9. Broken Patterns

1   Scranton, Roy, (2015) Learning to Die in the Anthropocene: Reflections on the End of a Civilization, City Light Books, p. 23
2   Titley, David, quoted by Roy Scranton, p. 81
3   Wiener, Ibid, p. 127
4   Wiener, Ibid, p. 34
5   Wiener, Ibid, p. 34
6   Knutson, Brian, (2018) Future Self-Continuity, in This Idea Is Brilliant, ed. John Brockman, Harper Perennial, p. 48

10. Feeling

1   Damasio, Antonio, (2010), Self Comes to Mind: Constructing the Conscious Brain, Vintage Books, p. 39
2   Dennett, Daniel C., (2017) From Bach to Bacteria and Back: The Evolution of Minds, Norton, p. 167
3   Pert, Candace B., (1997), Molecules of Emotion: The Science Behind Mind-Body Medicine, Scribner, p. 25
4   James, William, (1902) Varieties of Religious Experience, Penguin Classics, p. 53
5   Ibid, p. 101
6   Lightman, Alan, (2018), Searching for Stars on an Island in Maine, Pantheon Books, p. 6

310

7   Harari, 21 Lessons, Ibid, p. 161
8   Harari, Ibid, p. 163
9   Freud, Sigmund, Civilization and Its Discontents, p.13

11. Social Institutions
1   Scranton, Roy, Ibid. p. 2
2   Satinover, Jeffrey, (2001), The Quantum Brain, John Whiley and Sons, p. 49
3   Tooby and Cosmides, Ibid. p. 26
4   De Wall, Frans, (2013), The Bonobo and the Atheist: In Search of Humanism Among the Primates, Norton, p. 200
5   Blackmore, Susan, (1999), The Meme Machine, Oxford University Press, p. 3
6   Hofstadter, Douglas, (2007), I Am a Strange Loop, Basic Books, p. 213
7   Stoknes, Per Espen, (2015), What We Think About When We Try Not To Think About Global Warming: Toward a New Psychology of Climate Action, Chelsea Green, p. 7
8   Jamieson, Dale, (2014), Reason In a Dark Time, Oxford University Press, pp. 81-3
9   Marshall, George, (2014), Don't Even Think About It: Why Our Brains Are Wired to Ignore Climate Change, p. 19
10   De Wall, Ibid. pp. 228-235
11   Wohlleben, Peter, (2015), The Hidden Life of Trees: What They Feel, How They Communicate, Ludwig Verlag, Munich
12   Daly, Herman E., Ibid. pp. 31ff
13   Lebow, Victor, (1955), Price Competition in 1955, Journal of Retailing
14   Heisenberg, Ibid

12. Belonging
1   Berry, Fr. Thomas, (1999), The Great Work: Our Way Into the Future, Three Rivers Press, p. 82
2   Hilman, James, Ibid. p. xiii
3   Conway, Flo and Siegelman, Jim, (2005), Dark Hero of the Information Age: In Search of Norbert Wiener the Father of Cybernetics, p. 307
4   Berry, Wendell, (2002), The Art of the Commonplace: The Agrarian Essays of Wendell Berry, Shoemaker and Hoard, pp. 318-9
5   Gazzangia, Michael, (2018), The Schnitt, in This Idea Is Brilliant, ed. John Brockman, p. 157

13. Current Crises and The New Story
1   Campbell, Joseph, (1949), The Hero With a Thousand Faces, New World Library, p. 11
2   Berry, Fr. Thomas, (1988), The Dream of the Earth, Sierra Club Books, p. 123
3   Eliot, Thomas S., (1942), Little Gidding, Faber and Faber

4   Buchsel, Friedrich, "Eidolon", Vol. II TDNT, pp. 376-7
5   Leopold, Aldo, (1949), A Sand County Almanac, Oxford University Press, p. 239
6   Carroll, Sean, (2016), The Big Picture: On the Origins of Life, Meaning, and the Universe Itself, p.19
7   Kurzweil, Ibid. p. 279
8   Steiner, Christopher, (2012), Automate This: How Algorithms Took Over Our Markets, Our Jobs, and the World, Penguin Group, pp. 146-162
9   Ibid. pp. 178-182
10   Brown, Lester, (2006), Plan B 2.0: Rescuing a Planet Under Stress and a Civilization in Trouble, Norton, pp. 6 ff
11   Meadows, Donella, Randers, Jorgen, and Meadows, Dennis, (2004), Limits to Growth: The 30-Year Update, Chelsea Green
12   Cullather, Nick, (1999), Secret History: The CIA's Classified Account of Its Operations in Guatemala 1952-1954, Stanford University Press
13   Kronman, Ibid. pp. 823, 853-7
14   Charlton, Ibid. pp. 100 ff
15   De Waals, Ibid. pp. 215-6
16   Berry, Fr. Thomas, Ibid. p. xi
17   Keynes, John Maynard, Economic Possibilities for Our Grandchildren

14. The Last Enemy

1   Tuchman, Barbara, (1979), A Distant Mirror: The Calamitous Fourteenth Century. Knopf
2   Charlton, Ibid. p. 203
3   Bultmann, Rudolf, Thanatos, Vol. III TDNT, pp. 8-12
4   Pollan, Ibid. p. 109
5   Scranton, Ibid. pp. 21-2

Made in USA - Kendallville, IN
1208779 9781950381692